Y0-BQJ-828

L. German S. Zemskov (Eds.)

New Fluorinating Agents in Organic Synthesis

With Contributions by
L. A. Alekseeva, V. V. Bardin, L. S. Boguslavskaya, A. I. Burmakov,
N. N. Chuvatkin, G. G. Furin, B. V. Kunshenko, L. N. Markovskii,
F. M. Mukhametshin, V. E. Pashinnik, L. M. Yagupolskii,
Yu. L. Yagupolskii

Springer-Verlag Berlin Heidelberg New York
London Paris Tokyo Hong Kong

Dr. Lev Solomonovich German

Institute of Organoelement Compounds,
Vavilov St. 28, 117813 Moscow, USSR

Professor Dr. Stanislav Valerianovich Zemskov

Institute of Inorganic Chemistry, Pr. Lavrent'eva 3,
630090 Novosibirsk, USSR

ISBN 3-540-51160-1 Springer-Verlag Berlin Heidelberg New York
ISBN 0-387-51160-1 Springer-Verlag New York Berlin Heidelberg

Library of Congress Cataloging-in-Publication Data
New flourinating agents in organic synthesis / with contributions by
L. A. Alekseeva ... [et. al.] ; L. German, S. Zemskov, eds.
Includes index.
ISBN 0-387-51160-1 (U.S.)
1. Chemistry, Organic--Synthesis. 2. Flouridation.
I. Alekseeva, L. A. II. German, L. (Lev), 1931-. III. Zemskov, S. (Stanislav), 1934-.
QD262.N46   1989   547.2--dc20   89-10086 CIP

© Springer-Verlag Berlin Heidelberg 1989
Printed in Germany

Typesetting: Macmillan, Indien
Offsetprinting: Color-Druck Dorfi GmbH, Berlin; Bookbinding: Lüderitz & Bauer, Berlin
2151/3020-543210

In Memoriam

Professor S.V. Zemskov, one of the editors of the Russian edition of *New Fluorinating Agents in Organic Synthesis,* recently died. He was known as an outstanding researcher, a specialist in inorganic fluorine chemistry.
S.V. Zemskov was born in 1934. In 1960 he began work at the Institute of Inorganic Chemistry in Novosibirsk. Later he founded the laboratory of noble metal halide compounds. The scientific heritage of Professor Zemskov includes more than 150 publications and 30 patents, but the best memorial to his activities is the scientific school, his followers and this book.

Preface

Fluoroorganic compounds find extensive use in many fields of science, medicine, industry, and agriculture. However fluorination of organic compounds with elemental fluorine sometimes presents a serious problem. Along with classical fluorinating agents (HF, alkali metal fluorides, antimony fluorides, etc.), efficient agents for synthesis of fluoroorganic compounds are xenon fluorides, compounds with the O–F bond, N–F bond, and the higher fluorides of group V and VI elements. For most of these compounds, the preparative procedures have been developed only recently. Information on these synthetically interesting agents is scattered in many periodicals and patents. This book is intended to help researchers select the most suitable fluorinating agents and reaction conditions. For that purpose the book contains the necessary data on the new fluorinating agents illustrated by syntheses of some fluoroorganic compounds.

Recently new fields of application of sulfur tetrafluoride and fluorosulfuranes in fluoroorganic synthesis have been found. Therefore we considered it necessary to include reviews of applications of these fluorides. Due to space limitations, the mechanistic analysis of fluorination is incomplete; some interesting works dealing with this problem remained beyond the scope of this volume.

We hope that this book will stimulate further search for efficient fluorinating agents and applications of those already available.

July, 1989

L. German
S. Zemskov

Contents

1 Xenon Difluoride

Vadim Viktorovich Bardin[1] **and Yuri L'vovich Yagupolskii**[2]

[1] Institute of Organic Chemistry, Pr. Lavrent'eva 9, 630090 Novosibirsk, USSR
[2] Institute of Organic Chemistry, Murmanskaya St. 5, 252094 Kiev, USSR

Contents

1.1 Introduction

Three binar xenon fluorides are known (XeF_2, XeF_4 and XeF_6), among which the difluoride is the most studied. Relatively easy availability of XeF_2 allows it to be used to obtain organofluorine compounds in mild conditions. The peculiarity of xenon difluoride is its ability to carry out the "electrophilic" fluorination of organic compounds, i.e. addition of fluorine atoms proceeds as if an attacking species were "F". The reactivity of XeF_2 increases sharply in the presence of Lewis acids. At the same time under certain conditions (gas phase or irradiation) xenon difluoride may be a radical fluorinating agent.

This chapter considers the use of xenon difluoride in the synthesis of fluorine-containing organic and organoelement compounds of different types and demonstrates the synthetic utility of XeF_2. The mechanisms of these reactions, though also of interest, are not discussed here because of space limitations.

1.2   Synthesis and Some Properties of Xenon Difluoride. Interaction with Lewis Acids and Bases

The literature reports several methods for the preparation of xenon difluoride using various fluorinating agents [1,2]. Among them, the most popular is the photochemical method which involves the sunlight or the UV irradiation of a mixture of xenon and fluorine (1:1, mol) in a pyrex or quartz reactor [3]. This gives 99% pure xenon difluoride in amounts of several grams per day. To obtain larger amounts of XeF_2 there is a more convenient method based on the catalytic fluorination of xenon with chlorine trifluoride [4]. Besides, xenon difluoride formed in the process contains practically no higher xenon fluoride admixtures. As ClF_3 is commercially available and easier to handle than fluorine (see, e.g. [5]), Sect. 1.10 includes both preparations of XeF_2 [3,6]. Under normal conditions (20 °C, 1 atm) xenon difluoride is a colourless crystalline substance subliming upon heating (subl. p. 114 °C [2]). Purified XeF_2 melts at 141 °C (under pressure) [7]. The enthalpy of formation of xenon difluoride in the gas phase is -122 ± 3 kJ mol^{-1}, the heat of sublimation 51.5 ± 0.8 kJ mol^{-1}, the mean value of Xe–F bond thermochemical energy 140.6 kJ mol^{-1} [7]. The IR, Raman, ^{19}F and ^{129}Xe NMR, UV, X-ray and photoelectron spectral data for XeF_2 are reported in [1,2].

At 20 to 25 °C xenon difluoride is well-soluble in HF, BrF_3, IF_5, NOF·3HF, SO_2, SO_2FCl, CH_3CN, $(CF_3CO)_2O$; not so soluble in WF_6, MoF_6, CCl_4, diethyl ether, dimethyl sulfoxide; and slightly soluble in liquid ammonia, chlorine, bromine, trichlorofluoromethane, hydrocarbons. The solubilities of XeF_2 in some of these substances are given below, g in 1000 g of solvent (°C) [1,2,8].

HF	1329(12)	NOF·3HF	2755(17)	CH_3CN	167(0)	$SiCl_4$	1.9(22)
	1680(30)		5670(80)		309(21)		
IF_5	1538(20)	BrF_5	1893(25)	CCl_4	5(22)		

The acetonitrile solution of XeF_2 is stable at -10 °C, decomposes slowly at 20 °C, and quickly—when boiled. Xenon difluoride is moderately soluble in water (25 g in 1000 g of water), forming a molecular solution which decomposes with evolution of xenon (half-life period 7 h). In an alkaline medium, XeF_2 decomposes very quickly [2].

The mechanism of catalytic action of strong aprotic Lewis acids (SbF$_5$, NbF$_5$ etc.) involves ionization of the xenon difluoride molecule, leading to formation of fluoroxenonium salts $XeF^+ MF_n^-$ or $Xe_2F_3^+ MF_n^-$. These salts are strong oxidants: the electron affinity of XeF^+ is 10.8 ± 0.4 eV [2]. Less strong acceptors of the fluoride ion (HF, BF_3, BF_3OEt_2) polarise the XeF_2 molecule; an intermediate here is rather a complex of type $F–\overset{\delta+}{Xe} \ldots F \ldots \overset{\delta-}{BF_3}$ than the ion-structured salt.

Protic oxygen-containing acids (CF_3COOH, HSO_3F, HNO_3) react with xenon difluoride, forming unstable ethers FXeOY or $Xe(OY)_2$. Due to polarisation of bond $F\overset{\delta+}{Xe}–\overset{\delta-}{OY}$, the xenon atom has a partial positive charge; that is why strong protic acids are used as fluorination catalysts. At the same time, ethers $Xe(OY)_2$ are apt to decompose by the radical pathway, which leads to by-products [2,9].

Xenon difluoride practically does not react with fluoride ion donors at room temperature. Studies of the system XeF_2–CsF by the differential thermal analysis method have shown that only at the temperature of about 400 °C does slow disproportionation occur with the formation of $Xe + Cs_2XeF_8$ [10]. For this reason some cases of the catalytic effect of fluoride ion donors in the reactions of XeF_2 with organic substrates should be explained by interaction of F^- with these substrates rather than with XeF_2.

1.3   Fluorination at the Tetrahedric Carbon Atom

At room temperature xenon difluoride is insoluble in hexane and does not react with it [11]. Heating the XeF_2–hexane mixture to 80–90 °C in a sealed glass tube leads to its inflammation. The reaction products are HF, Xe, and tarry substances. Upon heating a mixture of C_6H_{14} and XeF_2 (5 : 1, mol) at 105 °C in a Teflon jacket inserted in a stainless steel reactor, the product is a mixture of 1-, 2- and 3-fluorohexanes (28 : 42 : 30) (total yield 15–20%).

$$C_6H_{14} + XeF_2 \xrightarrow[105\,°C, \ 2.5\,h]{} 1\text{-}FC_6H_{13} + 2\text{-}FC_6H_{13} + 3\text{-}FC_6H_{13}$$

In these conditions cyclohexane is converted to fluoro-cyclohexane in a 18% yield.

Heating of adamantane with xenon difluoride (105 °C, 70 min) leads to formation of a mixture of mono-, di-, and trifluoroadamantanes in a 18–20% yield. In this case, fluorine is almost exclusively substituted for the hydrogen atoms bound with the tertiary carbon, the main reaction product being 1-fluoroadamantane. In carbon disulfide, xenon difluoride fluorinates adamantane at temperatures as low as − 15 to 0 °C, giving 1-fluoroadamantane in yields of up to 35% [12].

Fluorination of tris(fluorosulfonyl)methane with xenon difluoride in CF_2Cl_2 leading to $CF(SO_2F)_3$ has been described, but no details of synthesis and yield of the product were given [13].

Substitution of hydrogen by fluorine is observed upon treatment of alkyl(aryl)methyl sulfides with xenon difluoride in acetonitrile [14–17]. Methyl-phenyl sulfide is transformed at room temperature to fluoromethylphenyl sulfide in a good yield and further to difluoromethylphenyl sulfide [14].

$$C_6H_5SCH_3 + XeF_2 \xrightarrow[25\,°C]{HF/CH_2Cl_2} C_6H_5SCH_2F \xrightarrow[25\,°C]{XeF_2/HF, \ CH_2Cl_2}$$

$$67\%$$

$$C_6H_5SCHF_2$$

$$58\%$$

Such fluorination will only take place if at the α-carbon atom there is at least one hydrogen atom. For example, phenylisopropyl sulfide reacts with XeF_2 (CH_3CN, $-10\,°C$, 0.5 h) giving phenyl(α-fluoroisopropyl) sulfide in a 90% yield, whereas fluorination of *tert*-butylphenyl sulfide and diphenyl sulfide produces the respective difluorosulfuranes in good yields [15]. It should be noted that phenyl(α-fluoroalkyl) sulfides (alkyl $\neq CH_3$) are unstable and with increase of temperature there increases the probability of HF elimination. This is vividly illustrated by the results of fluorination of phenylisopropyl sulfide by XeF_2 at different temperatures [15].

$$C_6H_5SCHMe_2 + XeF_2 \xrightarrow{MeCN}$$

$$\xrightarrow{-10\,°C,\ 30\ min} C_6H_5SCFMe_2 \quad 90\%$$

$$\xrightarrow{-5\,°C,\ 10\ min} C_6H_5SCFMe_2 + C_6H_5SC{=}CH_2 \ \underset{Me}{|}$$
$$\qquad\qquad\qquad\quad 60\% \qquad\qquad 40\%$$

$$\xrightarrow{0\,°C,\ 2\ min} C_6H_5SC{=}CH_2 \ \underset{Me}{|}$$
$$\qquad\qquad\qquad\quad 90\%$$

Apparently for the same reason it was impossible to isolate the products of fluorination of *cis*-2,6-diphenyltetrahydro-1-thio-4-pyrone and thiochroman-4-one. Instead of these compounds the reaction gives unsaturated products with good yields [14].

$$+ \ XeF_2 \xrightarrow{HF/CH_2Cl_2}$$
$$65\%$$

Yanzen and his co-workers [17] used xenon difluoride to obtain fluorine-containing derivatives of methionine and methionylglycine. It appeared that substitution of hydrogen by fluorine proceeds regiospecifically, and it is exclusively methyl at the sulfur atom that is fluorinated.

$$CH_3SCH_2CH_2\underset{NHCOR^2}{\underset{|}{CH}}COR^1 + XeF_2 \xrightarrow[-20\ to\ 20\,°C,\ -Xe,\ -HF]{MeCN} CH_2FSCH_2CH_2\underset{NHCOR^2}{\underset{|}{CH}}COR^1$$
$$\qquad\qquad\qquad\qquad\qquad\qquad\qquad\qquad\qquad\qquad\qquad\qquad 70–90\%$$

$R^1 = OMe$, $R^2 = CF_3$; $R^1 = 4\text{-}NO_2C_6H_4O\text{-}$, $R^2 = CF_3$; $R^1 = 4\text{-}NO_2C_6H_4O\text{-}$, $R^2 = t\text{-}Bu$; $R^1 = EtOCOCH_2NH$, $R^2 = PhCH_2O$

There are no products of fluorination of the α-methylene group, nor of deep fluorination of the methyl group.

By contrast with diorganyl sulfides, thiols are not fluorinated by xenon difluoride, but quantitatively oxidised to disulfides. Neither the structure of organic radical at the sulfur atom (R = Me, i-Pr, Ph) nor the ratio of thiol to XeF_2 affect the reaction course [15]. It is interesting that aliphatic alcohols (C_1–C_4) are oxidised by xenon difluoride to the respective aldehydes [18].

With carbon tetrachloride, xenon difluoride markedly reacts at 180 °C [8]. Introduction of catalytic amounts of antimony, niobium, tantalum fluorides and even SiO_2 leads to substitution of chlorine by fluorine at 20 to 30 °C with formation of chlorofluoromethanes.

$$CCl_4 + XeF_2 \xrightarrow[\text{20 to 30 °C}]{\text{Ct}} CFCl_3 + CF_2Cl_2 + CF_3Cl + CF_4$$

Ct		Yields %		
SbF_3	56	39	12	3
TaF_5	68	28	4	1
SiO_2	85	14	—	—

1.4 Fluorination of Alkenes and Alkynes

The first time the use of xenon difluoride for the synthesis of difluoroalkanes from alkenes was suggested was in Ref. [19]. This report immediately attracted the attention of researchers, as before that time there had been no method of direct fluorination of unsaturated hydrocarbons. Xenon difluoride reacts at room temperature with ethylene to form 1,2-difluoroethane, 1,1-difluoroethane, and 1,1,2-trifluoroethane [20].

$$CH_2=CH_2 + XeF_2 \xrightarrow[\text{25° C, 4 days}]{} CH_2F-CH_2F + CHF_2-CH_3 + CHF_2-CH_2F$$
$$ 45\% \qquad\qquad 35\% \qquad\quad 20\%$$

Fluorination of propylene by XeF_2 leads to formation of 1,2-difluoropropane (12%), 1,1-difluoropropane (46%), 2-fluoropropane (24%), and 1,1-difluoroethane (9%). The authors of [20] explain the complex composition of the reaction mixtures, and, particularly, the presence of geminal difluorides, by the radical character of the fluorination reaction. However it cannot be excluded that in the course of the reaction, isomerisation of vic-difluorides by elimination-addition of HF may occur. This is indicated by the formation of the product of addition of HF to propylene in the reaction of the latter with XeF_2.

Xenon difluoride fluorinates hexachlorocyclopentadiene relatively easily [21]. The reaction proceeds at room temperature in $CFCl_3$ without a catalyst and results in hexachlorodifluorocyclopentenes with a 93% yield. In a similar way XeF_2 reacts with 3-fluoropentachlorocyclopentadiene. The residual carbon–carbon double bond containing two vinyl chlorines is not fluorinated even in the presence of HF (60 h, 25 °C).

Fluorination of alkenes by xenon difluoride is markedly facilitated if the reactions are carried out in the presence of Lewis acids. Warming of a mixture of 1-decene, XeF_2, BF_3OEt_2 and CH_2Cl_2 from -78 to $20\,°C$ with subsequent stirring for 20 h leads to the formation of 1,1-difluorodecane (26%) and a mixture of 1-fluorodecene isomers (48%). However under the same conditions 2-fluoro-2,2-dinitroethyl vinyl ether is only converted to 1′,2′-difluoroethyl 2-fluoro-2,2-dinitroethyl ether [22].

$$(NO_2)_2CFCH_2OCH=CH_2 + XeF_2 \xrightarrow[\text{20 °C, 20 h}]{BF_3OEt_2/CH_2Cl_2}$$

$$\underset{79\%}{(NO_2)_2CFCH_2OCHFCH_2F}$$

Investigation of the reaction of xenon difluoride with aliphatic 1,3-dienes in the presence of BF_3OEt_2 has shown that under mild conditions this reaction proceeds with kinetic control, to form mainly the products of 1,2-addition of fluorine atoms [23].

$$CH_2=CH–CH=CH_2 + XeF_2 \xrightarrow[-10\,°C]{BF_3OEt_2/CH_2Cl_2} \underset{87\%}{CH_2FCHFCH=CH_2}$$

$$+ \underset{13\%}{CH_2FCH=CHCH_2F}$$

$$\underset{\overset{|}{H_3C}\ \overset{|}{CH_3}}{CH_2=C–C=CH_2} + XeF_2 \xrightarrow[25\,°C]{BF_3OEt_2/CH_2Cl_2} \underset{\overset{|}{H_3C}\ \underset{100\%}{\overset{|}{CH_3}}}{CH_2F–\ CFC=CH_2}$$

The reaction of xenon difluoride with phenyl-substituted alkenes has been studied in detail. Treatment of 1,1-diphenylethylene with XeF_2 at room temperature in the presence of HF quickly leads to formation of 1,2-difluoro-1,1-diphenylethane and a small amount of deoxybenzoine. Xenon difluoride reacts likewise with 1,1-diphenyl-2-fluoroethylene and 1,1-diphenyl-1-propene [24].

$$Ph_2C=CHR + XeF_2 \xrightarrow[\text{25 °C, 20 min}]{HF/CH_2Cl_2} \underset{65-95\%}{Ph_2CF–CHFR + PhCOCHRPh}$$

R = H, F, Me

If CF_3COOH is used as a catalyst, then instead of deoxybenzoine the reaction gives 1-trifluoroacetoxy-1,1-diphenyl-2-fluoroethane.

Cis- and *trans-*stilbenes are fluorinated in the same conditions with the formation of 1,2-difluoro-1,2-diphenylethane diastereomers [25]. In this case, fluorination of *trans-*stilbene proceeds stereoselectively, which indicates the preferential anti-addition of fluorine to alkene. Fluorination of *cis*-stilbene proceeds non-stereoselectively.

$$PhCH=CHPh + XeF_2 \xrightarrow[\text{20 °C, 30 min}]{\text{HF/CH}_2\text{Cl}_2} PhCHF–CHFPh$$

                                                                    90%

| *cis* | *erythro* : *threo* = 53 : 47 |
| *trans* | 62 : 38 |

As in the case of 1,1-diphenylethylene, fluorination of stilbenes by XeF_2 in the presence of CF_3COOH is accompanied by their fluorotrifluoroacetoxylation. In more detail the reaction of xenon difluoride with phenyl-substituted ethylenes in the presence of CF_3COOH has been analysed in [25–27].

Treatment of *trans*-1-phenylpropene and *trans*-stilbene with XeF_2 in CH_2Cl_2 catalysed by trifluoroacetic acid has been shown to lead to an about equimolar mixture of the products of fluorination and fluorotrifluoroacetoxylation of the starting alkenes. The same compounds are formed from the corresponding *cis*-isomers, but saturation of the double bond of *cis*-stilbene (2) (R = Ph) proceeds, by contrast with other alkenes, non-stereoselectively.

		erythro-3	*threo-3*	*erythro-4*	*threo-4*
1	R = Me	34	19	29	18
	R = Ph	30	18	35	17
2	R = Me	34	19	31	16
	R = Ph	26	24	26	24

It should be noted that fluorotrifluoroacetoxylation of compounds *1* and *2* (R = Me) proceeds regiospecifically.

In the reaction of xenon difluoride and CF_3COOH with styrene, apart from the products of fluorination and fluorotrifluoroacetoxylation of double bond, trifluoromethyl-containing compounds are formed [27].

$$PhCH=CH_2 + XeF_2 \xrightarrow[\text{25 °C, 1 h}]{\text{CF}_3\text{COOH}} PhCHF–CH_2F + PhCH–CH_2F$$

                                                                    OCOCF$_3$

                                       28%   29%

$$+ PhCHF–CH_2OCOCF_3 + PhCHF–CH_2CF_3 + PhCH–CH_2CF_3$$

                                                                    OCOCF$_3$

      8%   26%   9%

Thus, even small changes in the structure of phenyl-substituted alkenes affect the product ratio in their reaction with XeF_2 and CF_3COOH. At the same time, a property common for all the reactions of xenon difluoride with alkenes in the presence of trifluoroacetic acid is the formation of trifluoroacetyl derivatives.

$$XeF_2 + CF_3COOH \rightleftharpoons FXeOCOCF_3 + HF$$

$$PhCH(OCOCF_3)-CFR^1R^2$$

By contrast with hydrocarbon phenyl-substituted alkenes, the compounds containing halogen atoms at the double bond react slowly with xenon difluoride. Thus complete fluorination of 1,1-diphenyl-2-X-ethylene (X = F, Cl, Br) requires 25 h, with stirring at room temperature in the presence of HF. It is interesting that the products of 1,2-shift of the phenyl radical are not formed in this case [28] (cf. [24]).

$$Ph_2C=CHX + XeF_2 \xrightarrow[\text{25 °C, 25 min}]{\text{HF/CH}_2\text{Cl}_2} Ph_2CF-CHFX$$
$$50\%$$

$$X = F, Cl, Br$$

Introduction of substituents, such as Cl, Me, OMe, into position 3 or 4 of the aromatic ring of 1,1-diphenylethylene does not produce any marked effect upon the rate and stereochemistry of fluorination as compared with unsubstituted substrate [29].

Treatment of 1,1-diphenylethylene with xenon difluoride in the presence of bromine and HF leads to formation of 2-bromo-1-fluoro-1,1-diphenylethane instead of difluoroethane. Bromofluorination of 1,1-diphenylethylenes with halogen or methyl at position 2 proceeds in the same way. The reaction occurs quickly at room temperature and leads to the respective bromofluorides with high yields [30].

n=1 (94%)	79	:	21
n=2 (81%)	50	:	50
n=3 (91%)	35	:	65

According to [32], a change in the ratio of cis- to trans-difluorides from 1-phenylcyclopentene to 1-phenylcycloheptene is called for by steric reasons. Substitution of the phenyl group at the C=C bond by the 4-anisyl one does not affect the stereochemical outcome of fluorination of substituted cyclohexene, nor introduction of a bulky tert-butyl group into position 4 [33]. Treatment of

1-(4-anisyl)-4-*tert*-butylcyclohexene with xenon difluoride in the presence of chlorine or bromine source leads to the products of fluorohalogenation of the substrate [33].

In [34,35], reactions of xenon difluoride with acenaphthylene, indene, 1,2-dihydronaphthalene and 1,4-dihydronaphthalene have been studied. The products of fluorination of indene are *cis*- and *trans*-1,2-difluoroindans. Addition of fluorine atoms to 1,2-dihydronaphthalene occurs in a similar way.

$$n = 0\,(72\%)\qquad 30\quad :\quad 70$$
$$n = 1\,(70\%)\qquad 26\quad :\quad 74$$

Treatment of acenaphthylene with XeF_2 under the same conditions as for indene and 1,2-dihydronaphthalene leads to resinification of the reaction mixture. To avoid this, the reaction should be carried out in a diluted solution for 5 min. Yield of 1,2-difluoroacenaphthene is 68% (*cis* to *trans* = 16 to 84%).

Interaction of xenon difluoride with 1,4-dihydronaphthalene led to the formation of a mixture of naphthalene, 1- and 2-fluoronaphthalene and 2-fluorotetralin instead of the expected 2,3-difluorotetralin. This must be due to the instability of 2,3-difluorotetralin under the reaction conditions [35].

Treatment of alkene 5 with XeF_2 afforded the derivatives of 1,2-difluoro-1,2-dideoxyhexose [36].

Fluorination of compounds of type *5* with a different arrangement of AcO groups proceeds likewise. The main product in this case is always a difluoride containing both fluorine atoms in the *trans, trans*-position to the acetoxy group at position 3. As a solvent for this synthesis the authors of [36] recommend a mixture of ether and benzene, as in ether under the action of BF_3OEt_2, isomerisation of target products takes place.

A rather complex mixture of products is formed in the reaction of xenon difluoride with norbornene in the presence of acid catalysts [37–41]. The HF-catalysed fluorination of norbornene at room temperature leads to a mixture containing at least 7 components, and their yields practically do not depend on the solvent (CCl_4, $CHCl_3$ or CH_2Cl_2) [37].

Under milder conditions (-78 to $26°C$, 22 h) and in the presence of boron trifluoride etherate, the reaction affords a mixture containing only 2 main products -2-*exo*-5-*exo*-difluoronorbornane (*9*) and 2-*endo*-5-*exo*-difluoronor-bornane (*8*) (2:1) at a total yield of 51–76%. If the reaction is carried out at -46 to $-39°C$ for 75 min, the yields of compounds *8, 9* become almost equal (32 and 25% respectively), and the main product is 2-*exo*-7-*anti*-difluoronorbornane (*7*) (42%) [38]. The structure dependence of the products of fluorination of norbornene with XeF_2 on the solvent, temperature, reaction duration, nature of catalyst (HF, Py(HF)$_n$, BF$_3$, BF$_3$OEt$_2$, CF$_3$COOH, C$_6$F$_5$SH) and the routes of product isomerisation are analysed in more detail in [39,40]. If fluorination of norbornane with XeF_2 is carried out in $CFCl_3$ + MeCN without a catalyst and under UV irradiation, it gives, apart from the radical fluorination products *10* and *11*, a marked amount of compounds *12–14*. When the reaction is carried out in acetonitrile, the total yield of the latter increases [42].

It should be noted that the ion fluorination products 8 and 9, and the rearranged 2,5- and 2,7-difluoronorbornanes are not formed here.

Interaction of xenon difluoride with norbornadiene leads to 3-*endo*-5-*exo*-difluoronortricyclane, 3-*exo*-5-*exo*-difluoronortricyclane and 2-*exo*-7-*sin*-difluoro-5-norbornene [37,41,43]. Their relative yields change depending on the catalyst of fluorination.

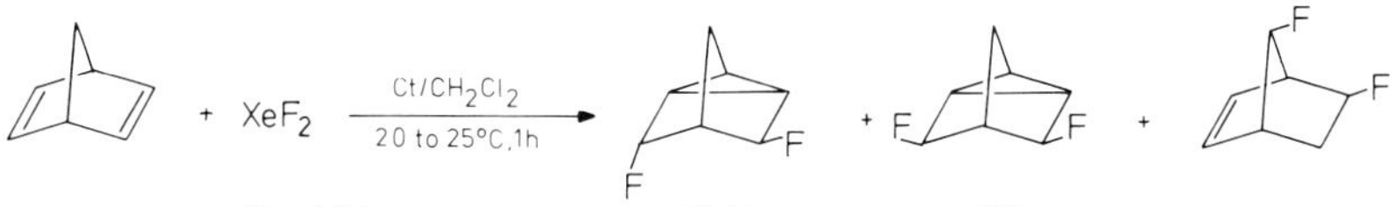

	50%	40%	10%
Ct = HF	50%	40%	10%
Ct = Py(HF)$_n$	27%	68%	5%
Ct = (P)—C$_5$H$_4$N(HF)$_n$	45%	44%	11%

Treatment of an alkene with xenon difluoride in the presence of methanol and acid catalyst (HF, BF$_3$) leads predominantly to methoxylation products [44]. Fluoroderivatives are formed in these conditions in low yields or not formed at all. Exclusions are, evidently, indene and dihydropyrane.

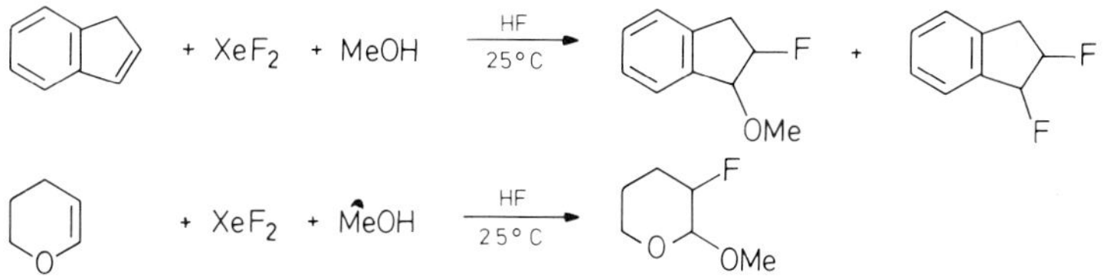

60% (33% *cis*, 67% *trans*)

At room temperature xenon difluoride fluorinates uracyl, giving 5-fluoro-uracyl, though the yield of the latter is small (10%) [45].

Literature also reports an unsuccessful attempt at fluorination of imidazo(1,2-b)pyridazine with xenon difluoride in CCl$_4$, CHCl$_3$ or CHBr$_3$ catalysed by HF or CF$_3$COOH [46]. The reaction product is 3-chloro-imidazo(1,2-b)pyridazine (or the corresponding brominated product); there were no fluorinated compounds in the reaction mixture. This reminds us once more that chlorinated methanes are not always inert in the reactions with xenon difluoride (see Sect. 1.3).

Xenon difluoride reacts with enols and their ethers. The products are α-fluorocarbonyl compounds. For example, treatment of 1-trimethylsilyloxy-cyclopentene or -hexene with an equimolar amount of XeF$_2$ leads to 2-fluorocyclopentanone and -hexanone respectively [47]. Acetates of the corresponding enols react in a similar way [48].

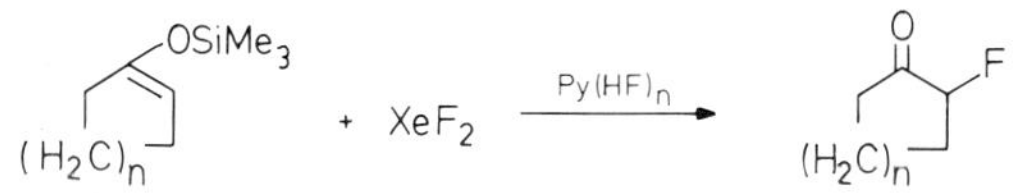

70-90%

The yield of 2-fluorocycloheptanone ($n = 3$) considerably decreases as a result of the side reaction of protodesilylation of enol ether which leads to cycloheptanone [47,48]. Fluorination of trimethylsilyl ethers of some steroid enols with XeF_2 proceeds without catalysts giving the least sterically hindered α-fluoroketones [49].

Enolised 1,3-diketones react with XeF_2 in the presence of protic acids, forming 2,2-difluorides. 2,4-Pentanedione reacts with xenon difluoride giving 3,3-difluoro-2,4-pentanedione (yield 70%). However fluorination of cyclic 1,3-diketones in the presence of HF is frequently complicated by side reactions and it is reasonable to use polymeric acids as a catalyst [48].

Fluorination of alkynes with xenon difluoride has been studied much less than fluorination of alkenes. In the absence of acid catalysts propyne has been noted [20] to react with XeF_2 more slowly than with propene, forming a complex mixture with 2,2-difluoropropane as the main component (yield 33%). Interaction of diphenylacetylene with xenon difluoride catalysed by an HF trace leads to 1,2-diphenyltetrafluoroethane [50]. This compound is also formed with a lack of XeF_2 (there also remains unreacted diphenylacetylene). Together with a relatively low reaction rate and the absence of difluorostilbenes in the reaction mixture this circumstance definitely shows that the reactivity of the triple carbon-carbon bond in the interaction with XeF_2 is much lower than that of the double bond.

$$PhC \equiv CR + 2XeF_2 \xrightarrow[25\,°C,\,6\,h]{HF/CH_2Cl_2} PhCF_2\text{–}CF_2R$$

$$R = Ph\,(50\%),\ Me\,(53\%),\ Pr\,(55\%)$$

In the presence of CF_3COOH, the interaction of diphenylacetylene with XeF_2 proceeds faster, but the yield of 1,2-diphenyltetrafluoroethane is only 4% [27]. The main reaction products are the trifluoroacetoxy- and trifluoromethyl-containing compounds.

1.5   Fluorination of Aromatic Compounds

Fluorination of benzene and some of its derivatives with xenon difluoride in gas phase is reported in [51]. Heating of benzene with an equimolar amount of XeF_2 in a nickel reactor gives fluorobenzene and a small amount of o- and p-difluorobenzenes. Nitrobenzene is converted at 120 °C to a mixture of o-, m- and p-fluoronitrobenzenes.

Fluorination of arenes in condensed phase has been studied more intensely. Benzene and its derivatives are fluorinated with xenon difluoride at room temperature and in the presence of HF to form predominantly monofluoro-arenes together with a small amount of difluorides. At the same time di- and polyphenyls are formed [52–56].

$$C_6H_5R + XeF_2 \xrightarrow[25\,°C]{HF/CCl_4} FC_6H_4R$$

R = H (68%), F (47%), Cl (66%), Me (32%), OMe (65%),
    CF$_3$ (76%), NO$_2$ (81%)

The isomer ratio corresponds with the orienting effect of the R substituents in the reactions with electrophilic agents. Thus, *ortho-*, *meta-* and *para-*chloro-fluorobenzenes are formed in yields 16.0, 3.2 and 46.3%, and fluorination of benzotrifluoride gives *meta-* and *para-*fluorobenzotrifluoride in yields 71.7 and 3.8% respectively [53].

Instead of HF, anhydrous hydrogen chloride may be used as a catalyst, but the yield of aryl fluorides is decreased by formation of aryl chlorides [55].

It should be noted that the catalytic activity of HCl is much lower than that of HF. For example, in HF (−75 °C) benzotrifluoride immediately reacts with XeF$_2$ to form fluorobenzotrifluorides, whereas in HCl (−75 °C) the compound remains unchanged for 2 days [55].

Interaction of polymethylbenzenes with xenon difluoride in the presence of HF leads to the products of fluorination of the aromatic ring, the methyl group remaining intact. Mesitylene reacts with two equivalents of XeF$_2$ to give 2,4-difluoromesitylene.

The HF-catalysed fluorination of 1,2,3-trimethylbenzene with XeF$_2$ (1:1, mol) gives a 1:2 mixture of 1-fluoro-3,4,5-trimethylbenzene and 1-fluoro-2,3,4-trimethylbenzene (yield 38%). Further fluorination of the mixture of these products leads to 1,2-difluoro-3,4,5-trimethylbenzene (yield 70%). While the reaction of 1,2,3-trimethylbenzene with XeF$_2$ (1:2, mol) gives a complex mixture of up to ten products, 1,2,4-trimethylbenzene is fluorinated by XeF$_2$ to 1-fluoro-2,4,5-trimethylbenzene.

Treatment of 1,2,4,5-tetramethylbenzene with an equimolar amount of XeF$_2$ (catalyst HF) leads to 1,4-difluorotetramethylbenzene. When the ratio of substrate to XeF$_2$ is increased to 1:2, the fluorination products are 1,4-difluoro-tetramethylbenzene and 1-fluoro-2,4,5-trimethylbenzene (3:2). Yields of these fluorine-containing arenes are small (approx. 7-9%) and together with these a considerable amount of tar is formed. At the same time, fluorination of polymethylbenzenes with xenon difluoride, catalysed by CF$_3$COOH, is complicated by the same processes of trifluoromethylation and trifluoroacetoxylation as the reactions of XeF$_2$ with olefins. In these reactions the trifluoromethyl group substitutes hydrogen atoms in the ring and the trifluoroacetoxy group – hydrogen atoms in the side chain [57].

Xenon difluoride reacts with hexamethylbenzene in the presence of HF forming pentamethylbenzyl fluoride. Upon substitution of HF with trifluoro-acetic acid there occurs trifluoroacetoxylation of hexamethylbenzene rather than fluorination, giving pentamethylbenzyl trifluoroacetate [58].

The reactions of xenon difluoride with benzocyclenes and related compounds have been studied [59–61]. Tetralin, 9,10-dihydroanthracene, and acenaphthene react with XeF$_2$ forming a mixture of monofluoroderivatives which contain fluorine in the aromatic ring. It should be noted that an isomer with the fluorine atom *ortho* to the methylene group is always formed in a small amount. A similar case is observed for the reaction of xenon difluoride with *o*-xylene. Indan [59] and fluorene [60] are almost completely fluorinated at the *meta*-position. Further fluorination of the resulting compounds leads to difluorides.

Fluorination of the aromatic ring of hydroxy- and alkoxyaromatic compounds by xenon difluoride proceeds easier than fluorination of arenes with such substituents as alkyl, F, Cl, CF_3, and NO_2. Thus conversion of anisole by XeF_2 to fluoroanisoles readily proceeds without catalyst, as well as fluorination of 1,2-dimethoxybenzene [62].

Phenol and pyrocatechol react with xenon difluoride in CH_2Cl_2 forming fluorophenols (47%) and 1,2-dihydroxy-4-fluorobenzene (38%) [62]. At the same time, upon treatment of hydroquinone and 4-*tert*-butyl-1,2-dihydroxybenzene with XeF_2 (CH_2Cl_2, 20 °C), instead of fluoroarenes one obtains benzoquinone and 4-*tert*-butyl-1,2-benzoquinone respectively [63].

It is interesting that in water phenol is oxidised by xenon difluoride to p-benzoquinone, presumably with the intermediate formation of hydroquinone [64, 65].

Dutch scientists [66] reported that the product ratio in the reaction of XeF_2 with 2-bromo-4,5-dimethylphenol (*15*) (catalyst BF_3OEt_2) strongly depends on the temperature and duration of the process. The nature of solvent (CH_2Cl_2, $CHCl_3$ or 1,2-$C_2H_4Cl_2$) is of no importance. For example, treatment of the substituted phenol *15* with an equimolar amount of XeF_2 at -65 to $-25\,°C$ leads to 2-bromo-4-fluoro-4,5-dimethylcyclohexa-2,5-dienone. But if the CH_2Cl_2 solution of phenol *15* and XeF_2 is stirred for 6 h at $-10\,°C$, then the usual treatment affords 3-bromo-5,6-dimethyl-1,2-benzoquinone (2.6%), 2-bromo-6-fluoro-4,5-dimethylphenol (5%) and 2-bromo-6-fluoro-3,4-dimethylphenol (7%).

Fluorination of the aromatic ring of L-4-hydroxy-3-methoxyphenylalanine proceeds more smoothly and gives L-6-fluoro-3,4-dihydroxyphenylalanine (after protodemethylation with HBr) [67]. This was used later in the synthesis of L-6-[^{18}F]dihydroxyphenylalanine [68].

Aromatic carboxylic acids react with xenon difluoride to form unstable ethers ArCOOXeF whose decomposition in excess benzene leads to aroyloxylation of the latter [69–72]. Phthalic acid is dehydrated by XeF_2 (accetonitrile, $20\,°C$) to phthalic anhydride in a 26% yield, whereas diphenyl-2,2'-dicarboxylic acid is transformed in the same conditions mainly to 3,4-benzocoumarine (yield 30%) and the respective anhydride is formed in very small amounts [73]. Saccharin reacts with XeF_2 in excess benzene ($60\,°C$, 48 h) to form N-phenyl-saccharin (22%), and phthalimide is converted to N-phenyl-phthalimide [74]. It should be stressed that no fluorine-containing compounds are formed in all these reactions.

Aniline and benzylamine are fluorinated by xenon difluoride so that the N–H bonds remain intact [75]. The products of the reaction of aniline with XeF_2 are o-, m- and p-fluoroanilines (yields 37, 3 and 11% respectively). The *ortho*-orientation is still more pronounced in fluorination of benzylamine.

$$C_6H_5NH_2 + XeF_2 \xrightarrow[-196\text{ to}-0\,°C]{\text{MeCN}} 2\text{-}FC_6H_4NH_2 + 3\text{-}FC_6H_4NH_2 + 4\text{-}FC_6H_4NH_2$$
$$\qquad\qquad\qquad\qquad\qquad\qquad\quad 37\% \qquad\qquad 3\% \qquad\qquad 11\%$$

$$C_6H_5CH_2NH_2 + XeF_2 \xrightarrow[-78\text{ to}-20\,°C]{CH_2Cl_2} 2\text{-}FC_6H_4CH_2NH_2 + 4\text{-}FC_6H_4CH_2NH_2$$
$$\qquad\qquad\qquad\qquad\qquad\qquad\qquad 40\% \qquad\qquad\qquad 2\%$$

It should be noted that these amines, like phenols, react with XeF_2 without catalyst.

Interaction of xenon difluoride with benzenesulphonyl chloride and related compounds in acetonitrile is limited to substitution of chlorine by fluorine in the functional group ($25\,°C$, 4 to 10 h) [76]. In methylphenylchlorosilane, substitution of chlorine by fluorine is faster (3 min), simultaneously fluorine is substituted for the hydrogen atom at silicon. A similar exchange reaction of XeF_2 occurs in the series of alkylchlorosilanes [77].

$$RSO_2Cl + XeF_2 \xrightarrow[\text{25 °C, 4 – 10 h}]{\text{MeCN}} RSO_2F$$
$$20\text{–}80\%$$

$$R = Me,\ Ph,\ C_6F_5,\ 4\text{-}CH_3C_6H_4$$

$$C_6H_5SiHClMe + XeF_2 \xrightarrow[\text{25 °C, 3 min}]{\text{MeCN}} C_6H_5SiF_2Me$$

The reactions of xenon difluoride with elementoaromatic compounds proceeding without oxidation of heteroatom in a side chain have been practically uninvestigated. Reutov and his co-authors [78,79] have shown that diaryl-mercury readily reacts with XeF_2 in mild conditions, forming aryl fluoride, arylmercuric fluoride, arene and diaryl. The reaction of xenon difluoride with dibenzylmercury and bis(phenylethynyl)mercury proceeds in a similar way.

$$R_2Hg + XeF_2 \xrightarrow[-45 \div 5\ °C]{CHCl_3} RF + RHgF + RH + R_2$$

$$R = 4\text{-}CH_3OC_6H_4,\ 4\text{-}Me_2NC_6H_4,\ 4\text{-}EtOCOC_6H_4,\ PhCH_2,\ PhC\equiv C$$

Because of complex composition of the reaction mixture and low yields, it is not practical to synthesize fluorides RF in this way.

Xenon difluoride is a convenient fluorinating agent for the synthesis of fluorine-containing polycyclic aromatic compounds many of which are difficult to obtain by conventional methods. But the reaction in the series of polycyclic compounds differs from fluorination of olefins and benzene derivatives. First, the reactivity of polycyclic arenes exceeds by far that of substituted benzenes and olefins, therefore in many cases the use of a catalyst is unnecessary. Second, in order to raise the yield of fluorinated arenes, the reagents should be mixed at a low temperature and the mixture slowly warmed to room temperature. Third, the reaction is carried out in very diluted solutions, otherwise the polymeric substance is formed in large amounts.

According to [80], naphthalene, anthracene and phenanthrene are fluorinated with xenon difluoride in the absence of a catalyst, giving a mixture of isomeric monofluoroarenes.

Later it has been found that fluorination of naphthalene gives in addition to 1- and 2-fluoronaphthalenes, 1,4-difluoronaphthalene (yields 46 and 15% respectively) [81]. The latter seems to be the product of fluorination of 1-fluoronaphthalene, but there is no direct evidence of it.

The reaction of xenon difluoride with phenanthrene was independently studied by three groups of researchers [80,82,83] who obtained similar results. Treatment of the diluted solution of phenanthrene ith XeF$_2$ in CH$_2$Cl$_2$ or CHCl$_3$ leads at first to 9-fluorophenanthrene (yield 33–60%). Raising the amount of XeF$_2$, addition of HF and prolonged keeping of the reaction mixture at room temperature raise the degree of fluorination, leading to 9,10-difluorophenan-threne, 9,9,10-trifluoro-9,10-dihydrophenanthrene and 9,9,10,10-tetrafluoro-9,10-dihydrophenanthrene. A special experiment has shown them to be the products of fluorination of 9-fluorophenanthrene [82].

An attempt to find in the reaction mixture 9,10-difluoro-9,10-dihydro-phenanthrene—the most probable primary product of the reaction of phenan-threne with XeF$_2$—was unsuccessful.

At $-78\,°$C, XeF$_2$ fluorinates pyrene and benzo[a]pyrene without catalyst. The general yield of monofluoroarenes is 20 to 30%. At the same time the reaction gives a marked amount of di- and trimeric products, which are isolated by TLC [84, 85].

This method allowed to obtain in one step the hardly available 1-fluoropyrene and 6-fluorobenzo[a]pyrene.

Perylene is transformed by XeF$_2$ (25 °C, 30 h) to 1-fluoroperylene, 3-fluoroperylene and difluoroperylene of unknown structure, but conversion of perylene in this reaction is small (40%) [86].

Only one work reports the results of studies on the reaction of xenon difluoride with pyridine and quinoline derivatives [75]. By contrast with benzene derivatives, pyridine easily reacts with XeF$_2$ without a catalyst, forming 2-fluoropyridine, 3-fluoropyridine and 2,6-difluoropyridine.

The ease of this reaction is the more surprising in that pyridine reacts with electrophilic agents in much more rigid conditions than benzene.

There are no data on the reaction of XeF_2 with quinoline, but 8-hydroxy-quinoline at 5 to 25 °C is readily converted to 5-fluoro-8-hydroxyquinoline (yield 35%) and a mixture of other compounds whose structure was not determined [75].

1.6 Fluorination of Polyfluoroaromatic Compounds

Polyfluoroaromatic compounds, like their hydrocarbon analogues, react with xenon difluoride in the presence of Lewis acids, but the fluorination proceeds by regiospecific 1,4-addition of fluorine even in the case of pentafluorobenzene and deuteropentafluorobenzene [87–91]. Position of the R substituent in the product of fluorination of substituted pentafluorobenzene C_6F_5R depends on its nature: with R = H, D, Cl, Br, C_6F_5 the reaction gives 1-R-heptafluoro-1,4-cyclohexadienes, with R = OAlk—a mixture of 1-R- and 3-R-heptafluoro-1,4-cyclohexadienes. The latter are unstable in the reaction conditions and are converted to hexafluoro-2,5-cyclodien-1-one.

Fluorination of octafluoronaphthalene with xenon difluoride catalysed by boron trifluoride or tungsten hexafluoride leads to perfluoro-1,2- and -1,4-dihydronaphthalene in the ratio of 1 to 7 [90, 92]. At the same time, the single product of fluorination of 2-methoxy- and 2-ethoxyheptafluoronaphthalene is 2-alkoxyperfluoro-1,4-dihydronaphthalene [92].

The above fluorination reactions of polyfluoroaromatic compounds with XeF_2 proceed at room temperature and are catalysed by HF, BF_3 or WF_6. However attempts to fluorinate pentafluorobenzonitrile and nitropentafluoro-benzene in the presence of BF_3 at 25 °C failed [89], though in the HF solution of xenon difluoride, nitropentafluorobenzene is transformed to 1-nitroheptafluoro-1,4-cyclohexadiene (yield 32%) [90]. Much stronger fluorinating agents are the fluoroxenonium salts, for example $XeF^+SbF_6^-$ [88, 90, 91]. Under the action of these fluorooxidants transformation of polyfluoroaromatic compounds to 1,4-cyclohexadiene derivatives proceeds even at -70 to $-80\,°C$.

R=F, NO₂

Boron trifluoride and HF have no catalytic effect in these conditions. Hence, upon fluorination of polyfluoroaromatic compounds by XeF_2, the catalytic activity of Lewis acids decreases in the series $SbF_5 > HF > BF_3$ [90].

Literature reports no data on fluorination of polyfluoroarenes containing the hydrocarbon alkyl groups. There is a description of the transformation of pentafluorophenol by XeF_2 to compound $C_{12}F_{10}O_2$, to which the structure of $C_6F_5OOC_6F_5$ was initially assigned [93]. Later it was established that actually the reaction gives a mixture of isomeric perfluorophenoxycyclohexadienones, and the reaction carried out in HF yields in addition perfluoro-2,5-cyclohexadien-1-one [94].

Trimethylsilylperfluoroarenes react with xenon difluoride in the presence of potassium or caesium fluorides with the aryl–silicon bond cleavage. It is interesting that in the absence of a fluoride ion source the reaction fails to proceed [95].

$$Ar_FSiMe_3 + XeF_2 \xrightarrow[20\ °C]{MeCN,\ CsF} Ar_FH + Ar_F\text{–}Ar_F + FSiMe_3 + Xe$$

$$R_F = C_6F_5,\ 4\text{-}CF_3C_6F_4,\ 4\text{-}C_5F_4N$$

A key intermediate here seems to be the perfluoroaryl radical. This is confirmed by the formation of Ar_FBr and $Ar_FC_6H_5$ when the reaction is conducted in the presence of $CHBr_3$ or benzene respectively, and by the observed chemically induced polarization of ^{19}F nuclei in C_6F_5H and $C_6F_5\text{–}C_6F_5$ [95].

1.7   Oxidative Fluorination of Organoelement Compounds

Since the synthesis of xenon difluoride, one of the most attractive prospects was its application for the synthesis of organoelement compounds where an element is in one of the highest stages of oxidation. In 1968, Bartlett [96] showed the possibility of oxidative fluorination of iodine to IF_5, and of sulfur dioxide—to SO_2F_2 by XeF_2 in the presence of HF or BF_3. Subsequently the Australian chemists [97] carried out an extensive investigation of xenon difluoride as a reagent for the synthesis of inorganic fluorides. As the starting compounds they used elements, their oxides, halides and even carbonyls. The authors of [97] used the 10^{-2} to 1 M HF solution of xenon difluoride, the reactions proceeded at room temperature for several minutes. Below are given the highest oxidation degrees obtained for different elements using XeF_2.

Group	II	III	IV	V	VI
Element	Hg(2+)	Tl(3+)	Sn(4+)	As(5+), Sb(5+), Nb(5+)	S(6+), Cr(4+), Mo(6+), W(6+)

Group	VII	VIII
Element	I(5+), Mn(3+), Re(6+)	Co(3+), Ni(2+), Ru(5+), Rh(4+), Os(6+)

Among the compounds which were synthesized, it is worthwhile to mention carbonyl fluorides of tungsten, molybdenum, rhenium and ruthenium as representing the class of compounds unavailable by other methods.

The oxidative fluorination of organoelement compounds by xenon difluoride was studied by investigating the reactions of phosphorus-, sulfur-, selenium-, tellurium-, antimony-, arsenic- and iodoorganic derivatives. The carbon-element bond remains intact in these reactions.

Transformation of phosphines by XeF_2 to difluorophosphoranes proceeds in mild conditions (low temperature, MeCN or CH_2Cl_2 as solvent) and with high yields. Fluorination of the phosphine containing the hydroxy group and synthesized from diphenylphosphine and hexafluoroacetone occurs exclusively at the phosphorus atom [98].

$$(C_6H_5)_2PC(OH)(CF_3)_2 + XeF_2 \xrightarrow[\substack{-63\ °C,\ 2\ h \\ -23\ °C,\ 1\ h}]{MeCN - CFCl_3} (C_6H_5)_2PF_2C(OH)(CF_3)_2$$

The P–H bond is stable towards XeF_2, therefore alkyl- or arylphosphines are readily transformed to difluorophosphoranes [99].

$$R_2PH + XeF_2 \xrightarrow[-15\ °C]{MeCN} R_2PF_2H$$

$$RPH_2 + XeF_2 \xrightarrow{MeCN} RPH_2F_2 \quad (R = Ph, NCCH_2CH_2)$$
$$> 90\%$$

Chlorine-containing phenylphosphines react with XeF_2 forming fluorophosphoranes and chlorine; the intermediate formation of chlorofluorophosphoranes was not observed.

$$Ph_nPCl_{3-n} + XeF_2 \xrightarrow[-10\ to\ 0\ °C]{} Ph_nPF_{5-n}$$

In the case of bis(*tert*-butyl)chlorophosphine the reaction gives a mixture of the products of fluorination of methyl groups [99]. However the reaction with methyldiphenylphosphine or *o*-tolyldi(ethyl)phosphine leads exclusively to

difluorophosphoranes with the alkyl groups remaining intact [100]. Triphenyl-
and trimethyldifluorophosphorane may be obtained by fluorination of R_3P in a
90–100% yield [99,100].

$$R_3P + XeF_2 \rightarrow R_3PF_2 \quad (R = Me, Ph)$$

In compounds containing the bonds P–N, P–O, P–S, transformation of the
difluorophosphino group to the tetrafluorophosphoranyl one by XeF_2 is
complicated by a slight elimination of PF_5 and formation of the products
containing the P=X fragment, where X=N, O, S. Thus the reaction of
bis(difluorophosphino)methylamine with xenon difluoride proceeds with evolution of PF_5 and formation of difluorophosphino(tetrafluorophosphoranyl)methylamine, which further reacts with XeF_2 forming the iminophos-
phorane dimer and PF_5 [102].

$$MeN(PF_2)_2 + XeF_2 \xrightarrow[-PF_5]{} MeN(PF_2)PF_4 \xrightarrow[-PF_5]{XeF_2} (MeNPF_3)_2$$

$$(PF_2)_2X + XeF_2 \rightarrow F_3PX + PF_5 \quad (X = O, S)$$

Investigation of the oxidative fluorination of sulfur-containing compounds
was started by Zupan [14] (see Sect. 3). The sulfides containing no α-hydrogen
atoms were shown to transform smoothly to difluorosulfuranes, whereas the
presence of methyl, ethyl, isopropyl or benzyl groups at the sulfur atom prevents
isolation of difluorosulfuranes. In this case the reaction gives α-fluorine-
containing sulfides – the product of further transformations of compounds with
the SF_2 group [15].

$$Ph_2S + XeF_2 \rightarrow Ph_2SF_2$$

$$PhSCMe_3 + XeF_2 \rightarrow PhSF_2CMe_3$$

$$RSCHR_2' + XeF_2 \rightarrow RSCFR_2' + HF$$

The fluorination is carried out in the acetonitrile solution at low temperatures. In
these conditions formation of sulfur (VI) is not observed. But the reaction of
xenon difluoride with diphenyl sulfide in the $MeCN–CFCl_3$ mixture in the
presence of the catalytic amount of HF with subsequent hydrolysis leads to
diphenyl sulfone [103,104]. Duration of the reaction is 15 h; after 1 h, mainly
diphenyl sulfoxide is isolated.

$$Ph_2S + XeF_2 \xrightarrow[15\,h]{} [Ph_2SF_4] \xrightarrow{H_2O} Ph_2SO_2$$

The literature contains a detailed report on the fluorination of dimethyl
sulfide by XeF_2 giving Me_2SF_2 [16]. Interaction of the reagents at 20 °C without
solvent leads to an explosion but upon dilution with $CFCl_3$ it yields $MeSCH_2F$
which reacts with excess dimethyl sulfide, giving $[Me_2SCH_2SMe]^+[F(HF)_n]^-$.
In the anhydrous hydrogen fluoride solution, the reaction product is
$[Me_2SF]^+[F(HF)_n]^-$; which was strictly proved by the ^{19}F NMR data and an

alternative synthesis of Me_2SF_2 from Me_2S and AgF_2.

$$Me_2S + XeF_2 \xrightarrow[-HF]{} MeSCH_2F \xrightarrow{Me_2S-HF} [Me_2SCH_2SMe]^+[F(HF)_n]^-$$

$$Me_2S + XeF_2 + HF \rightarrow [Me_2SF]^+[F(HF)_n]^-$$

Substitution of one methyl group by the trifluoromethyl one in dimethyl sulfide does not alter the character of the end product of fluorination, the product being CF_3SCH_2F. But bis(trifluoromethyl) sulfide is oxidised to difluorosulfurane [101].

$$CF_3SCH_3 + XeF_2 \rightarrow CF_3SCH_2F$$

$$CF_3SCF_3 + XeF_2 \rightarrow (CF_3)_2SF_2$$

Despite the fact that aryltrifluoromethyl sulfides do not add chlorine, these compounds were transformed with the help of XeF_2 to aryltrifluoromethylsulfur difluoride in the presence of the catalytic amount of HF [105].

$$4\text{-}XC_6H_4SCF_3 + XeF_2 \xrightarrow{40 \text{ to } 50\,^\circ C} 4\text{-}XC_6H_4SF_2CF_3$$

$$X = H, Cl, NO_2$$

Thiols and thiophenols are oxidised by XeF_2 to disulfides [15] with evolution of HF, which allows to use C_6F_5SH as a catalyst in fluorination of organic compounds with XeF_2. But the disulfides can also react in rigid conditions with xenon difluoride. As shown by one of the authors, melting of decafluorodiphenyldisulfide with XeF_2 at 85 °C in a sealed tube for 3 h leads to the formation of a mixture of $C_6F_5SF_3$ and C_6F_5SOF, as determined by ^{19}F NMR.

$$2RSH + XeF_2 \rightarrow RS\text{-}SR + 2HF$$

$$C_6F_5SSC_6F_5 + XeF_2 \rightarrow 2C_6F_5SF_3 \xrightarrow{H_2O} C_6F_5SOF$$

Xenon difluoride is a very mild and highly selective fluorinating agent for compounds with the sulfur–nitrogen bond. Studies of the methyleneamidosulfenyl system $(CF_3)_2C=N\text{-}SX-$ have shown that at 20 °C, the oxidative fluorination proceeds with exclusive formation of the derivatives of $S(IV)$, whereas upon addition of BF_3 as a catalyst, the products containing $S(VI)$ are formed. Thus hexafluoroisopropylidenimino(trifluoromethyl) sulfide is quantitatively converted in mild conditions to sulfoximide [106].

$$(CF_3)_2C=N\text{-}SCF_3 + XeF_2 \rightarrow (CF_3)_2CF\text{-}N=SFCF_3$$

Hexafluoroisopropylidenimidosulfenyl isocyanate is transformed at room temperature to products *16* and *17* in high yields, corresponding to 1,3- and 1,5-difluorination of the substrate, the fluorination proceeding without catalyst

presence of BF_3 etherate, giving diarylbromonium salts whose structure was proved by an independent synthesis.

$$ArBr + XeF_2 \rightarrow [ArBrF_2] \xrightarrow{Ar'H,\ BF_3} [ArBrAr']^+\ BF_4^-$$

Xenon difluoride is a mild selective agent for the oxidative fluorination of the aromatic compounds of Sb(III). Thus diphenylfluorostibine and triphenylstibine are converted by XeF_2 to diphenyltrifluoroantimony and triphenyldifluoroantimony respectively [126].

$$Ph_2SbF + XeF_2 \xrightarrow[25\,°C]{CH_2Cl_2} Ph_2SbF_3$$
$$98\%$$

$$Ph_3Sb + XeF_2 \rightarrow Ph_3SbF_2$$
$$95\%$$

The methyl derivatives of As(III) are also readily transformed to the corresponding difluorides [100, 101].

$$CH_3AsR_2 + XeF_2 \rightarrow CH_3AsR_2F_2 \quad (R = CH_3,\ C_6H_5)$$

It is interesting that the trifluoroacetate group bonded with the antimony atom is substituted under the action of XeF_2 by the fluorine atom [126].

$$Ph_2SbOCOCF_3 + XeF_2 \xrightarrow[25\,°C]{CH_2Cl_2} Ph_2SbF_3$$
$$75\%$$

Up to now there have been no extensive studies on the reaction of xenon difluoride with organoelement compounds containing the metal-carbon bond (except organomercury compounds, see Sect.1.5). The possibility of this bond being retained in oxidative fluorination would open up new prospects of synthesis of organometallic fluorides in the highest stages of oxidation. But the question of the metal-carbon bond stability in the presence of XeF_2 should be decided for each metal individually. Recently it has been found [127] that the methylene chloride solutions of organoaluminium compounds Et_3Al, i-Pr_3Al, i-Pr_2AlH, Et_2AlCl, and $EtAlCl_2$ treated by xenon difluoride intensively luminesce. The oxidation is believed by the authors to proceed with cleavage of the Al–C bond. Among the stable products of the reaction was Et_2AlF, as shown by the mass-spectral data. The detailed investigation of the reactions of alkylaluminium compounds with xenon difluoride in toluene confirmed these results and led to a number of alkylaluminium fluorides R_2AlF ($R = Et$, i-Bu, EtO) at 80–97% yields [128].

It should be noted that the oxidative fluorination of organoelement compounds is a promising preparative method for the synthesis of compounds of new types, and work in this field, owing to the enhanced availability of xenon difluoride, will be intensively developed.

1.8 Other Fluorination Reactions

Some fluorination reactions of organic compounds with xenon difluoride may not be assigned to any of the above types. These involve, for example, the reactions of fluorodecarboxylation of alkanoic, arylalkanoic and aryloxyacetic acids with XeF_2 [129,130]. Alkanoic acids react with xenon difluoride in CH_2Cl_2 or $CHCl_3$ giving fluoroalkanes. The reaction may be performed at room temperature and usually requires 8–16 h. Pure products are obtained simply by washing the reaction mixture with sodium carbonate solution followed by removal of the solvent.

$$RCOOH + XeF_2 \xrightarrow[-Xe, -HF, -CO_2]{} RF$$

$R = n\text{-}C_9H_{19}\,(54\%),\ n\text{-}C_{15}H_{31}\,(62\%),\ PhCH_2\,(76\%),$
$\quad PhCH_2CH_2\,(76\%),\ PhCH_2CH_2CH_2\,(60\%),\ Ph_2CHCH_2\,(63\%),$
$\quad PhOCH_2\,(64\%),\ 2,4\text{-}Cl_2C_6H_3OCH_2\,(84\%),$
$\quad CH_2BrCH_2CH_2\,(91\%)$

It should be noted that no aromatic fluorination proceeds here.
Dicarboxylic acids are converted easily to difluoroalkanes.

$$PhCH_2CH(COOH)_2 + 2XeF_2 \rightarrow PhCH_2CHF_2$$
$$68\%$$

Tertiary carboxylic acids are decarboxylated easily with XeF_2. For example, 1-adamantanoic acid is converted to 1-fluoroadamantane (82% yield), and fluorotriphenylmethane (65%) is obtained from triphenylacetic acid and XeF_2. However 3-phenylbicyclo[1.1.1]pentan-1-oic acid reacts with XeF_2 to form a dimer, and no fluoro product is obtained here.

Several secondary carboxylic acids tested for fluorodecarboxylation give only small amount of fluoro products.

$$(CH_3)_2CHCOOH + XeF_2 \rightarrow (CH_3)_2CHF$$
$$10\%$$
$$(CH_2)_5CHCOOH + XeF_2 \rightarrow (CH_2)_5CHF$$
$$3\%$$

Benzoic acid gives small amount of benzoyl fluoride (see Sect. 1.5).

The ketone function of levulinic acid is unaffected in the reaction with XeF_2, and 4-fluoro-2-butanone is obtained (82% yield). But amino acids and cholic acid give only the unreacted starting material.

In view of these limitations, fluorodecarboxylation of carboxylic acids by xenon difluoride may be recommended as a convenient method for preparing small amounts of pure fluoroalkanes and arylfluoroalkanes.

Imides of perfluorinated succinic and glutaric acids react with XeF_2 forming the respective *N*-fluoroimides in good yields [131].

n=2(55-65%), 3(50-60%)

The fluorination is carried out at 0 to 20 °C without solvent or at 0 °C in CF_2Cl_2 (24 h, sealed tube). It is interesting that such NH-acid as bis(sulfuryl fluoride)amine reacts with XeF_2 in the same conditions, forming $FXeN(SO_2F)_2$ [132].

The reaction of xenon difluoride with 1,1-dinitroethane potassium salt has been described [133]. One of the products of the reaction is 1-fluoro-1,1-dinitroethane, but its yield is small and strongly depends on the reaction conditions (solvent, temperature, reagent ratio).

Finally, the reaction of xenon difluoride with the CF_3 radicals should be mentioned, which afforded bis(trifluoromethyl)xenon, $Xe(CF_3)_2$—the first and as yet the only organoxenon compound [134].

1.9   Conclusions

We give here some practical recommendations on fluorination of organic compounds with xenon difluoride.

In the first works on fluorination with XeF_2 the reactions were carried out in metal vacuum lines. Later it appeared that XeF_2 may be handled in open systems. Most convenient for fluorination are the reactors of quartz, polychlorotrifluoroethylene or polytetrafluoroethylene; the pyrex reactor may also be used, but only in cases when it does not react with a catalyst. Radical fluorination under heating or irradiation should be carried out in a quartz reactor.

The recommended solvents for fluorination in the condensed phase are CH_2Cl_2, $CHCl_3$, CH_3CN, SO_2FCl, and ether. The most widely used of them is dichloromethane, though xenon difluoride is slightly soluble in it. Acetonitrile dissolves XeF_2 very well but reacts with it in the presence of Lewis acids. The most convenient solvent for the fluorinations catalysed by such acids as SbF_5, is SO_2FCl. It easily dissolves xenon difluoride, most organic compounds and the compounds initiating the reactions with XeF_2. Being a low-boiling liquid (b.p. 8 °C), SO_2FCl is easily isolated from the reaction mixture. The preparation of SO_2FCl from sulfuryl chloride and ammonium fluoride is described in [135].

In the cases when xenon difluoride does not directly react with a substrate, the catalytic amount of Lewis acid is added (10 to 20 mol %). The most efficient of these is SbF_5, but usually used in practice are boron trifluoride, BF_3OEt_2, HF and CF_3COOH. Fluorination in the presence of trifluoroacetic acid proceeds with trifluoroacetoxylation and trifluoromethylation of a substrate, therefore CF_3COOH as the catalyst of fluorination is not recommended. A reasonably active fluorination catalyst is tungsten hexafluoride, but it is not as easily available as BF_3 or BF_3 etherate.

The Lewis acid-catalysed fluorination of alkenes, alkynes, aromatic and polyfluoroaromatic compounds with xenon difluoride proceeds as a rule at room temperature. The reaction of XeF_2 with aniline, alkoxybenzenes, polycyclic arenes and pyridines are carried out in milder conditions (low temperature, diluted solutions, no catalyst). This increases the yields of fluorine-containing products by suppressing polymerisation of substrates. The presence in the molecule of aromatic compound of such substituents as F, Cl, Br, alkyl, hydroxyalkyl, CN, COOR, NO_2 does not hinder the fluorine introduction in the aromatic ring. But phenols, aryl iodides, arylcarboxylic acids, 1-arylolefins and 1-arylalkynes treated with XeF_2 do not form the respective aryl fluorides, so for the successful synthesis of the latter these functional groups must be protected.

1.10   Preparations

1. Xenon difluoride [4]

A copper load was charged with 5 g of MnF_3 distributed in a thin layer, and placed into a heated horizontal copper reactor. The heated zone should exceed the load length by 30 to 50%, and a trap under cooling should be fixed as near the heated zone as possible to avoid concentration of XeF_2 in the reactor. The temperature in the reactor was kept at 320 to 345 °C. A mixture of xenon and chlorine trifluoride (1:0.7 to 1.0, mol) was passed at the rate of 3.8 to 7.2 l h^{-1}. Xenon difluoride condensed in the trap as colourless crystals. Yield 60–80%.

2. Xenon difluoride [6].

A 2 l Pyrex glass flask was filled with the gaseous mixture of 350 ml of xenon and 374 ml of fluorine. The total pressure in the flask was 724 mm at 25°C. The flask was kept for 3 weeks at room temperature exposed to daylight. The formation of fine crystals was noticed on the second day of standing. The initial rate of formation of XeF_2 was about 35 mg per day. The total amount of product obtained during 3 weeks varied from 0.5 to 0.75 g.

3. cis, trans-1,2-Difluoroacenaphthene [35]

In a Kel-F reactor was placed 1 mmol of acenaphthylene in 50 ml of CH_2Cl_2, then anhydrous HF (traces) was added and 1 mmol of XeF_2 was introduced with stirring. The colourless solution became dark blue, and xenon evolved. After 5 min the mixture was diluted with 15 ml of CH_2Cl_2, washed with 5% solution of sodium carbonate (10 ml), with water and dried. The solvent was distilled off and the residue subjected to TLC (silica gel, eluent–petroleum ether (40 to 60 °C)) to obtain cis-1,2-difluoroacenaphthene (8%, m.p. 103 to 105 °C) and trans-1,2-difluoroacenaphthene (60%, m.p. 41 to 43 °C).

In a similar way, *cis*- and *trans*-1,2-difluorotetralins were prepared (yields 10 and 60% respectively) [35], *cis*- and *trans*-1,2-difluoro-1-phenylcyclopentane (20 and 74%), -cyclohexane (40 and 40%), and -cycloheptane (59 and 32%) [32], with the only difference that the starting alkene was dissolved in 5 ml of dichloromethane and fluorination was carried out for 20–30 min.

4. 1-*R*-Heptafluoro-1,4-cyclohexadienes

A [89] In a Kel-F reactor was placed a solution of 2 mmol of substituted pentafluorobenzene C_6F_5R in 4 ml of dichloromethane, 2 mmol of xenon difluoride, then boron trifluoride was introduced with stirring. The mixture was stirred for 10–30 min at room temperature, then treated as described above, and the residue, after distilling off the solvent, was purified by the preparative GLC to give 1-*R*-heptafluoro-1,4-cyclohexadiene with approx. 80% yield (R = H, F, Cl, Br, C_6F_5).

B [90] A solution of 10 mmol of hexafluorobenzene in 20 ml of HF and 20 ml of SO_2FCl was placed in a Kel-F reactor, then cooled to -70 to $-80\,°C$, whereupon 11 mmol of $XeFSbF_6$ was added to it in portions, with stirring (preparation of $XeFSbF_6$ see in [136]). The mixture was further stirred for 25–30 min, then poured onto ice cooled with liquid nitrogen, the organic layer was separated, washed with water and dried. SO_2FCl (b.p. 8 °C) was distilled off to give perfluoro-1,4-cyclohexadiene in a 91% yield. In a similar way, 1-nitroheptafluoro-1,4-cyclohexadiene and 1-nitro-4-trifluoromethylhexafluoro-1,4-cyclohexadiene were prepared from nitropentafluorobenzene and 4-nitro-heptafluorotoluene respectively (yields 13 to 30%).

5. Fluorination of anisole [62]

A solution of 37 mmol of anisole in 12 ml of dichloromethane was placed in a Kel-F reactor and degassed at $-196\,°C/5 \cdot 10^{-6}$ mm Hg. Then it was added to 12.2 mmol of XeF_2 placed in a similar reactor. The mixture was slowly heated to evolution of xenon, which was accompanied by darkening of the solution. The reaction usually proceeds at -10 to $25\,°C$ for several minutes. Distillation gave fluoroanisoles in a 71.5% yield ($o:m:p = 10:1:8$).

6. (Difluoroidodo)benzene [120]

In a quartz reactor connected with an apparatus for measuring the amount of xenon evolved, in a dry nitrogen atmosphere at 20 °C, was placed 10 mmol of iodobenzene and 11 mmol of xenon difluoride. After the theoretical amount of xenon had evolved, the product was distilled in vacuum, in a quartz apparatus. Yield of (difluoroiodo)benzene 80%, m.p. 30 to 40 °C, b.p. 93 °C/0.2 mm Hg.

In a similar way, (difluoroiodo)pentafluorobenzene was obtained from iodopentafluorobenzene, but to finish the reaction, the mixture was heated to 40 °C. Yield 82%, m.p. 48 to 49 °C, b.p. 90 °C/0.2 mm Hg.

7. General procedure for fluorodecarboxylation of carboxylic acids [130]

To a solution of carboxylic acid (1 mmol) in 15 ml of dichloromethane contained in a polyethylene bottle, was added xenon difluoride (1 mmol, 170 mg). The solution was stirred magnetically at 22 °C for 10 h during which the colourless solution became slightly yellow. The resulting mixture was washed with 3% sodium bicarbonate (50 ml) solution. The organic solution was dried ($MgSO_4$) and concentrated to yield the pure product.

1.11   References

1. Neiding AB, Sokolov VB (1974) Usp. Khim. 43:2146; (1975) Chem. Abs. 82:67498
2. Sladky F (1973) In: Gutmann V (ed) Main group elements group VII and noble gases. Butterworth, London, p 2
3. Williamson SM (1968) Inorg. Synth. 11:147
4. Mit'kin VN, Zemskov SV (1981) Izv. Akad. Nauk SSSR. Neorg. mater. 17:1897; (1982) Chem. Abs. 96:58511
5. Hudlicky M (1976) Comprehensive chemistry of organic fluorine compounds. Wiley, New York
6. Streng LV, Streng AG (1965) Inorg. Chem. 4:1370
7. Pepekin VI, Lebedev YuA, Apin AYu (1969) Zh. Fiz. Khim. 43:1564; (1969) Chem. Abs. 71:74947
8. Legasov VA, Marinin AC (1972) Zh. Neorg. Khim. 27:2408; (1973) Chem. Abs. 78:51974
9. Eisenberg M, Des Marteau D (1980) Inorg. Nucl. Chem. Lett. 6:29
10. Kiselev YuM, Goryachenkov SA, Martynenko LI, Spitsyn VI (1984) Dokl. Akad. Nauk SSSR 278:881; (1985) Chem. Abs. 102:105129
11. Zajc B, Zupan M (1986) Bull. Chem. Soc. Jap. 59:1659
12. Podkhalyuzin AT, Nazarova MP (1975) Zh. Org. Khim. 11:1568; (1975) Chem. Abs. 83:96539
13. Kloter G, Pritzkov H, Seppelt K (1980) Angew. Chem. 92:954
14. Zupan M (1976) J. Fluor. Chem. 8:305
15. Marat RK, Janzen AF (1977) Can. J. Chem. 55:3031
16. Forster AM, Downs AJ: J. Chem. Soc. Dalton Trans. 1984:2827
17. Janzen AF, Wang PM, Lemire AE (1983) J. Fluor. Chem. 22:557
18. Fehér I, Sempteg M (1970) Magy. Kem. Foly. 76:141; (1970) Chem. Abs. 73:24576
19. Shien TC, Chernick CL (1964) J. Amer. Chem. Soc. 86:5021
20. Shien TC, Feit E, Chernick CL, Yang N (1970) J. Org. Chem. 35:4020
21. Paprott G, Lentz D, Seppelt K (1984) Chem. Ber. 117:1153
22. Shackelford SA, McGuire RR, Pflug JL: Tetrahedron Lett. 1977:363
23. Shellhamer DF, Conner RJ, Richardson RF, Heasley VL (1984) J. Org. Chem. 49:5015
24. Zupan M, Pollak A: J. Chem. Soc. Chem. Commun. 1973:845
25. Zupan M, Pollak A: Tetrahedron Lett. 1974:1015
26. Zupan M, Pollak A (1977) Tetrahedron 33:1071
27. Gregorcic A, Zupan M (1979) J. Org. Chem. 44:4120
28. Gregorcic A, Zupan M (1979) J. Org. Chem. 44:1255
29. Zupan M, Pollak A (1976) J. Org. Chem. 41:4002
30. Stavber S, Zupan M (1977) J. Fluor. Chem. 10:271
31. Shellhamer DF, Ragains ML, Gipe BT, Heasley VL, Heasley GE (1982) J. Fluor. Chem. 20:13
32. Zupan M, Sket B (1978) J. Org. Chem. 43:698
33. Gregorcic A, Zupan M (1984) J. Org. Chem. 49:333
34. Zupan M, Pollak A (1977) J. Org. Chem. 42:1559
35. Sket B, Zupan M: J. Chem. Soc. Perkin Trans. I. 1977:2169
36. Korytnyk W, Valentekovie-Horvath S, Petrie CR (1982) Tetrahedron 38:2547
37. Zupan M, Gregorcic A, Pollak A (1977) J. Org. Chem. 42:1562
38. Shackelford SA: Tetrahedron Lett. 1977:4215
39. Shackelford SA (1979) J. Org. Chem. 44:3489

40. Gregorcic A, Zupan M (1980) Bull. Chem. Soc. Jap. 53:1085
41. Gregorcic A, Zupan M (1984) J. Fluor. Chem. 24:291
42. Hildreth RA, Druelinger ML, Shackelford SA (1982) Tetrahedron Lett. 23:1059
43. Gregorcic A, Zupan M (1977) Tetrahedron 23:3243
44. Shellhamer DF, Curtis CM, Dunham RH, Hollingsworth DR, Ragains ML, Richardson RE, Heasley VL, Shackelford SA, Heasley GE (1980) J. Org. Chem. 50:2751
45. Yurasova TI (1974) Zh. Obshch. Khim. 44:956; (1974) Chem. Abs. 81:25624
46. Zupan M, Pollak A (1976) J. Fluor. Chem. 8:275
47. Cantrell GL, Filler R (1985) J. Fluor. Chem. 27:35
48. Zajc B, Zupan M (1982) J. Org. Chem. 47:573
49. Tsushima T, Kawada K, Tsuji T (1982) Tetrahedron Lett. 23:1165
50. Zupan M, Pollak A (1974) J. Org. Chem. 39:2646
51. MacKenzie D, Fajer J (1970) J. Amer. Chem. Soc. 92:4994
52. Shaw MJ, Human HH, Filler R (1969) J. Amer. Chem. Soc. 91:1563
53. Shaw MJ, Human HH, Filler R (1970) J. Amer. Chem. Soc. 92:6498
54. US Pat 3833581 (1974); (1974) C. A. 81:135684
55. Shaw MJ, Human HH, Filler R (1970) J. Org. Chem. 36:2917
56. Turkina MYa, Gragerov IP (1975) Zh. Org. Khim. 11:340
57. Stavber S, Zupan M (1983) J. Org. Chem. 48:2223
58. Zupan M (1976) Chimia 30:305
59. Sket B, Zupan M (1978) J. Org. Chem. 43:835
60. Sket B, Zupan M (1981) Bull. Chem. Soc. Jap. 54:279
61. Filler R (1978) Isr. J. Chem. 17:71
62. Anand SR, Quaterman LA, Human HH, Migliorese KG, Filler R (1975) J. Org. Chem. 40:807
63. Zupan M, Pollak A (1976) J. Fluor. Chem. 7:443
64. Goncharov AA, Kozlov YuN, Purmal AP (1977) Zh. Fiz. Khim. 51:2939; (1978) Chem. Abs. 88:36967
65. Goncharov AA, Kozlov YuN (1978) Zh. Fiz. Khim. 52:945; (1978) Chem. Abs. 89:5703
66. Koundstaal H, Olieman C (1981) Rec. Trav. Chim. Pays Bas. 100:246
67. Firnau G, Chirakal R, Sood S, Garnett ES (1980) Can. J. Chem. 58:1449
68. Firnau G, Chirakal R, Sood S, Garnett ES (1981) J. Label. Compounds Radio-pharm. 18:7
69. Nikolenko LN, Shustov LD, Bocharova TN, Yurasova TI, Legasov VA (1972) Dokl. Akad. Nauk SSSR 204:1369; (1972) Chem. Abs. 77:101059
70. Bocharova TN, Marchenkova NG, Shustov LD, Prokof'eva TYu, Nikolenko LN (1973) Zh. Obshch. Khim. 43:1325; (1973) Chem. Abs. 79:65953
71. Shustov LD, Tel'kovskaya TD, Nikolenko LN (1974) Zh. Obshch. Khim. 44:2564; (1975) Chem. Abs. 82:97795
72. Shustov LD, Tel'kovskaya TD, Nikolenko LN (1975) Zh. Org. Khim. 11:2137; (1976) Chem. Abs. 84:30028
73. Shustov LD, Semenova MN, Nikolenko LN (1978) Zh. Obshch. Khim. 48:1903; (1979) Chem. Abs. 90:22739
74. Shustov LD, Nikolenko LN (1983) Zh. Obshch. Khim. 53:2408; (1984) Chem. Abs. 100:139008
75. Anand SR, Filler R (1976) J. Fluor. Chem. 7:179

76. Volkova SA, Sinyutina ZM, Nikolenko LN (1974) Zh. Obshch. Khim. 44:2592; (1975) Chem. Abs. 82:31070
77. Gibson JA, Janzen AF (1971) Can. J. Chem. 49:2168
78. Butin KP, Kiselev YuM, Magdesieva TV, Reutov OA (1982) Izv. Akad. Nauk SSSR. Ser. Khim.: 716; (1982) Chem. Abs. 97:39056
79. Butin KP, Kiselev YuM, Magdesieva TV, Reutov OA (1982) J. Organometal. Chem. 235:127
80. Anand SP, Quaterman LA, Christian PA, Human HH, Filler R (1975) J. Org. Chem. 40:3796
81. Rabinovitz M, Agranat I, Selig H, Lin CH (1977) J. Fluor. Chem. 10:159
82. Zupan M, Pollak A (1975) J. Org. Chem. 40:3794
83. Agranat I, Rabinovitz M, Selig H, Lin CH: Chem. Lett. 1975:1271
84. Bergmann ED, Selig H, Lin CH, Rabinovitz M, Agranat I (1975) J. Org. Chem. 40:3793
85. Agranat I, Rabinovitz M, Selig H, Lin CH (1976) Experientia 32:417
86. Stephenson MT, Shine HI (1981) J. Org. Chem. 46:3139
87. Stavber S, Zupan M: J. Chem. Soc. Chem. Commun. 1978:969
88. Bardin VV, Furin GG, Yakobson GG (1980) In: Vsesoyuznaya Konferenciya pamyati AE Favorskogo, Feb 1980. Leningrad, p 110
89. Stavber S, Zupan M (1981) J. Org. Chem. 46:300
90. Bardin VV, Furin GG, Yakobson GG (1982) Zh. Org. Khim. 18:604; (1982) Chem. Abs. 97:72000
91. Yakobson GG, Bardin VV, Furin GG (1983) In: The Third Regular meeting of Soviet-Japanese fluorine chemists. Tokyo, p 14.1
92. Zajc B, Zupan M (1982) Bull. Chem. Soc. Jap. 55:1617
93. Nikolenko LN, Yurasova TI, Man'ko AA (1970) Zh. Obshch. Khim. 40:938; (1970) Chem. Abs. 73:34956
94. Avramenko AA, Bardin VV, Karelin AI, Krasil'nikov VA, Tushin PP, Furin GG, Yakobson GG (1985) Zh. Org. Khim. 21:822; (1985) Chem. Abs. 103:141551
95. Bardin VV, Stennikova IV, Furin GG, Leshina TV, Yakobson GG (1988) Zh. Obshch. Khim. 58:2580
96. Bartlett N, Sladky F: J. Chem. Soc. Chem. Commun. 1968:1046
97. Burns RC, MacLeod ID, O'Donnell TA, Peel TE, Phillips KA, Waugh AB (1977) J. Inorg. Nucl. Chem. 39:1737
98. Janzen AF, Vaidya OC (1973) Can. J. Chem. 51:1136
99. Gibson JA, Marat RK, Janzen AF (1975) Can. J. Chem. 53:3044
100. Alam K, Janzen AF (1987) J. Fluor. Chem. 36:179
101. Forster AM, Downs AI (1985) Polyhedron 4:1625
102. Cowley AH, Chung Yi, Lee R (1979) Inorg. Chem. 18:60
103. Zupan M, Zajc B: J. Chem. Soc. Perkin Trans. I 1978:965
104. Gregorcic A, Zajc B, Zupan M 1979) Phosphorus Sulfur 6:107
105. Yagupol'skii YuL, Savina TI (1979) Zh. Org. Khim. 15:438; (1979) C. A. 90:203618
106. Varwig I, Mews R: J. Chem. Res. (Synops) 1977:245
107. Steinbeisser H, Mews R (1981) J. Fluor. Chem. 17:505
108. Geisel M, Mews R (1982) Chem. Ber. 115:2135
109. Leidinger W, Sundermeyer F (1982) Chem. Ber. 115:2892
110. Bergman I, Engman L (1981) J. Amer. Chem. Soc. 103:2715
111. Alam K, Janzen AF (1985) J. Fluor. Chem. 27:467

112. Naumann D, Herberg S (1982) J. Fluor. Chem. 19:205
113. Klein G, Naumann D (1985) J. Fluor. Chem. 30:259
114. Gibson JF, Janzen AF: J. Chem. Soc. Chem. Commun. 1973:739
115. Zupan M, Pollak A: Tetrahedron Lett. 1975:3525
116. Zupan M, Pollak A (1976) J. Org. Chem. 41:2179
117. Zupan M, Pollak A: J. Chem. Soc. Perkin Trans. I 1976:1745
118. Zupan M: Synthesis 1976:473
119. Stavber S, Zupan M (1978) J. Fluor. Chem. 12:307
120. Maletina II, Orda VV, Aleinikov NN, Korsunskii BL, Yagupol'skii LM (1976) Zh.
     Org. Khim. 12:1371; (1976) Chem. Abs. 85:77778
121. Zupan M, Pollak A (1976) J. Fluor. Chem. 7:445
122. Gregorcic A, Zupan M (1977) Bull. Chem. Soc. Jap. 50:517
123. Zupan M, Pollak A: J. Chem. Soc. Chem. Commun. 1975:715
124. Zupan M: J. Chem. Soc. Chem. Commun. 1977:266
125. Nesmeyanov AN, Lisichkina IN, Tolstaya TP (1978) Dokl. Akad. Nauk SSSR
     246:1463; (1978) Chem. Abs. 88:169693
126. Yagupol'skii LM, Popov VI, Kondratenko NV, Korsunskii BL, Aleinikov VP
     (1975) Zh. Org. Khim. 11:459; (1975) Chem. Abs. 82:171158
127. Bulgakov RG, Maistrenko GYa, Tolstikov GA, Yakovlev VN, Kazakov VP: Izv.
     Akad. Nauk SSSR. Ser. Khim. 1984:2644; (1985) Chem. Abs. 102:122365
128. Bulgakov RG, Yakovlev VN, Maistrenko GYa, Tolstikov GA: Izv. Akad. Nauk
     SSSR. Ser. Khim. 1986:490; (1987) Chem. Abs. 106:84699
129. Patrick TB, Johri KK, White DH (1983) J. Org. Chem. 48:4158
130. Patrick TB, Johri KK, White DH, Bertrand WS, Mokhtar R, Kilbourn MR, Welck
     MJ (1986) Can. J. Chem. 64:138
131. Yagupol'skii YuL, Savina TI (1981) Zh. Org. Khim. 17:1330; (1981) Chem. Abs.
     95:168500
132. LeBlond R, DesMarteau DD: J. Chem. Soc. Chem. Commun. 1974:555
133. Celinskii IV, Mel'nikov AA, Varyagina LG, Trubizyn AE (1985) Zh. Org. Khim.
     21:2490. Celinskii IV, Mel'nikov AA, Trubizyn AE (1987) Zh. Org. Khim. 23:1657
134. Turbini L, Aikman R, Lagow R (1979) J. Amer. Chem. Soc. 101:5833
135. Woyski M (1950) J. Amer. Chem. Soc. 72:919
136. Gillespie R, Netzer A, Schrobilgen G (1974) Inorg. Chem. 13:1455

2 Some "Electrophilic" Fluorination Agents

Georgii Georgievich Furin

Institute of Organic Chemistry, Pr. Lavrent'eva 9, 630090 Novosibirsk, USSR

Contents

2.1 Introduction

Advances in studies of the properties of fluoroorganic compounds continuously extend the field of their practical applications. At first, only for fluorinated polymers and freons could a use be found, whereas now the range of applications of fluoroorganic compounds is so wide and they are so varied that this has become a separate field of chemistry [1,2].

The new field was based on the classic fluorination methods worked out at the beginning of the 1960s. Their advantages were described in a monograph [3], which had a great effect on the further development of fluoroorganic chemistry. The progress in this field, which was achieved in the 1980s was largely determined by the fundamental approach of the pioneers of fluoroorganic chemistry [4]. Today, the main demerit of fluoroorganic compounds from the practical viewpoint is their high cost owing to the high toxicity and corrosion activity of fluorine and hydrogen fluoride, and the difficulty of introducing the required number of fluorine atoms into the strictly definite positions of organic molecules. With the advance of science and technology these demerits are being gradually overcome. This is due to the unique physical and chemical properties of some polyfluorinated compounds, which in their turn stimulate the development of new synthetic methods and improvement of the existing technologies. Thus the possibility of perfluorinated tertiary amines, ethers, and paraffins being used as the components of artificial blood [5–7] led to significant advances in the development of electrochemical and gas-phase fluorination methods.

Diethyl phenylmalonate and its sodium salt have been shown [18] to undergo the substitutive fluorination by perfluoropiperidine and perfluorinated polymers 4. The fluorination product is diethyl phenyl-2-fluoromalonate, 1 being a more efficient fluorinating agent than the N–F-containing fluoropolymers 4.

$$PhCH(COOEt)_2 \longrightarrow \begin{cases} \xrightarrow{1} & PhCF(COOEt)_2 \\ & 68\% \\ \xrightarrow{4} & PhCF(COOEt)_2 \\ & 17\text{-}23\% \end{cases}$$

4: $-(CF_2-CF)_n$ m = 0,1   R = 4

An efficient fluorinating agent is N-fluoro-2-pyridone obtained by the reaction of fluorine with 2-trimethylsiloxypyridine [19].

This agent may be used to obtain α-fluorocarbonyl compounds [20].

$$RCH=C-R \xrightarrow[2)H_2O]{1)5} RCHFCR$$

R = Ph (11–33%), RR = cyclo-C_4H_8 (36–44%)

N-Fluoropyridone (5) allows us to carry out the regiospecific fluorodemetallation of Grignard reagents and the sodium salts of diethyl malonate derivatives [19,20].

$$RMgBr + 5 \rightarrow RF$$

R = Ph(15%), cyclo-C_6H_{11}(11%), 2-C_8H_{17} (5%)

$$R\bar{C}(COOEt)Na^+ + 5 \rightarrow RCF(COOEt)_2$$

R = H (9%), Me(17%), Ph (39%)

In the latter case, with R = PhCH$_2$, the fluorinated ester is easily decarboxylated to give ethyl 2-fluoro-3-phenylpropanoate (30–39%) [19].

Regretfully, there are few examples of fluorination by N-fluoro-2-pyridone, so it is difficult to evaluate its synthetic utility.

An interesting compound is formed upon treatment of pyridine with fluorine at a low temperature [21–23]. It is suggested to have the structure of N-fluoropyridinium fluoride [23]

This compound was used for the fluorination of uracyl and some chloroolefins [23,24]. However, the adduct is of little use because of its violent decomposition above −2 °C.

An important achievement of the Japanese scientists is the synthesis of N-fluoropyridinium triflate which, as opposed to the fluoride, is safe, and stable against hydrolysis [24–26]. This method is rather universal, it allows one to obtain a wide range of substituted N-fluoropyridinium triflates and study their fluorinating ability [25,26]. It is interesting to note that introduction of various substituents into the pyridine ring allows variation of the fluorinating ability of the salts to be made.

Pathway A:

R=H(6,80%), 2-Me(60%), 2,6-Me$_2$(73%), 2,4,6-Me$_3$(7,49%), 2-OMe(73%), 2,6-(COOMe)$_2$(9, 72%)

Pathway B:

R=H(6,78%), 3-Cl(79%), 3,5-Cl$_2$(8,55%), 3-COOMe(69%), 2,6-(COOMe)$_2$(9,68%)

The substituted N-fluoropyridinium triflates were used for substitution of hydrogen by fluorine in aromatic compounds [26–28]. Thus benzene was transformed to fluorobenzene (yield 56%, triflate 9), N-(carboxyethyl)aniline gave 2- and 4-fluoro-N-(carboxyethyl)anilines (yields 60 and 27% respectively, triflate 6), phenylurethane also gave the products of fluorination at position 2 (47%) and 4 (32%) of the aromatic ring (triflate 9). In the case of fluorination of

Table 1. (*Continued*)

Compound	Reagent	Product	Yield (%)
Me₂CHCH₂ Et C=C Li H	D	Me₂CHCH₂ Et C=C F H	83
O Bu Li	D	O Bu F	74
=Li (steroid)	D	=F (steroid)	80

A: *N*-Fluoro-*N*-neopentyl-*p*-toluenesulfonamide; B: *N*-Fluoro-*N*-*tert*-butyl-*p*-toluene-sulfonamide; C: *N*-Fluoro-*N*-*exo*-2-norbornyl-*p*-toluenesulfonamide; D: *N*-Fluoro-*N*-*tert*-butylbenzenesulfonamide

Table 2. Fluorination of aromatic compounds with $(CF_3SO_2)_2NF$ (CDCl₃, 22 °C) [38]

Compound[a]	Time (h)	Products (%)
Nitrobenzene[b]	12	no reaction
Acetophenone	12	no reaction
Chlorobenzene	24	no reaction
Benzene[b]	18	fluorobenzene, 50
Toluene[b]	10	2-fluorotoluene, 74; 3-fluorotolune, 4; 4-fluorotoluene, 22
Anisole[b]	2	2-fluoroanisole, 69; 4-fluoroanisole, 24; polyfluoroanisole, 7
Phenol[c]	12	2-fluorophenol, 60; 4-fluorophenol, 40
m-Cresol[c]	12	2-fluoro-5-methylphenol, 44; 4-fluoro-3-methylphenol, 56
p-Cresol[c]	12	2-fluoro-4-methylphenol, 80; other, 20
m-Xylene[c]	12	2-fluoro-5-methyltoluene, 2-fluoro-3-methyltoluene (1:2 mixture)
Naphthalene[c]	12	1-fluoronapthalene, 80; 2-fluoronapthalene, 7; other, 13

[a] ArH: $(CF_3SO_2)_2NF = (\geqslant 2): 1$, mol
[b] Without solvent
[c] An NMR yield (a percent of fluorinated products)

CF_3OF [35–37].

$$RSO_2NHR' \xrightarrow{F_2-N_2} RSO_2NFR'$$

$R' = R = $ alkyl (C_{1-30}), cyclo-alkyl (C_{3-30}), aryl, aralkyl
$R = $ 4-tolyl, $R' = $ Me, t-Bu, cyclo-C_6H_{11}, t-$BuCH_2$, 2-norbornyl
$R = $ Bu, $R' = t$-$BuCH_2$

Table 1 represents some examples of fluorination by these compounds.

It is worthwhile paying attention to the high yields (71–88%) of the fluorodemetallation products of alkenyllithium. This reaction shows the stereo- and regiospecificity of fluorination and the absence of polymers, due to which it may be recommended as a preparative method for the synthesis of fluoroolefins.

$$R_1R_2C=CLiR_3 + PhSO_2NF(t\text{-}Bu) \to R_1R_2C=CFR_3$$

The enhanced electron-accepting ability of substituents at the nitrogen atom in nitrogen-fluorine containing compounds raises the reactivity of the latter. These fluorides are obtained by the reaction of 1% F_2 in N_2 with bis(perfluoroalkylsulfonyl)imides in $CFCl_3$ ($-78\,°C$) with 60–96% yields [38].

Among N-fluoro-bis(perfluoroalkylsulfonyl)imides, the reactions of $(CF_3SO_2)_2NF$ are the most extensively studied ones. This fluoride reacts with benzene, toluene, xylene, and other aromatic compounds with electron-donating substituents, but is inert with chlorobenzene and nitrobenzene (Table 2). The orientation in this fluorination is formally that of an electrophilic attack, though the reaction gives an unusually large amount of *ortho*-isomer. There are no polyfluorinated compounds among the products, which seems to result from the low nucleophilicity of monofluorinated arenes as compared with their precursors. The selectivity of fluorination observed here is especially important, because the partially fluorinated benzenes are used for the synthesis of biologically active compounds, though their use is restricted by the absence of simple and convenient methods of synthesis [39].

Literature contains no data on the reaction of perfluorinated N-fluorosulfonylamides with unsaturated compounds.

2.3 Alkaline Metal Fluorooxysulfates

Since Appelman reported, in 1979, the first synthesis of caesium fluorooxysulfate by the reaction of fluorine with caesium sulfate in water [40], there has been constant growth of interest in this reagent. As a rule, caesium and, more rarely, rubidium salts are used. These salts are potential oxidants and fluorinating agents. The standard electrode potential of $SO_4F^--HSO_4^-$ is 2.5 V [41]. Detailed structural analysis of these salts led to a conclusion that caesium fluorooxysulfate is a hypofluorite of ionic structure. It is already clear that such salts are synthetically and analytically convenient in organic and inorganic

Table 3. Relative reactivities of C_6H_5R toward SO_4F^- (MeCN, 25 °C) [41]

R	Reactivity
NO_2	0.02
CN	0.07
COOMe	0.17
F	0.55
H	1.00
Ph	11
Me	13–29
OMe	190
OH	440

the electron-donating substituents to increase the fluorination rate, and the electron-accepting ones—to decrease it [48]. The ρ^+ value found from the data of Table 3 equals -3.5. This indicates that this reagent is more selective than the elementary fluorine for which the ρ^+ value is -2.45.

The effect of catalysts: HF, H_2SO_4, CF_3SO_3H, HSO_3F, BF_3, and SbF_5–HSO_3F in the fluorinations of toluene, nitrobenzene, and naphthalene by fluorooxysulfates has been studied in detail [43]. The acid catalysts accelerate the reaction, and the efficiency of the catalyst parallels the H_0 function of the respective acid. In particular, in the fluorination of naphthalene with $CsSO_4F$, H_2SO_4 was found to be a more efficient catalyst than HSO_3F. It is worth giving here some considerations on the mechanism of fluorination of aromatic compounds by caesium fluorooxysulfate. It should be noted that no detailed mechanistic studies have been carried out, therefore these considerations are rather speculative.

The key intermediate is supposed to be complex A. Elimination of SO_4^{2-} produces the benzenonium cation, whereas elimination of SO_4^- and F^- leads to radical cation C, implying the occurrence of the one-electron oxidation of the aromatic compound. The benzenonium ion B gives the product of fluorination of the aromatic substrate, whereas radical cation C undergoes further transformation, forming a lot of by-products. The latter prevail in the case of non-activated or deactivated aromatic substrates.

Protic and Lewis acids promote the transformation of intermediate A to ion B, thus raising the yield of the products of fluorination in the benzene ring. The leaving group here is HSO_4^-, and in the case of the BF_3 catalyst—$BF_3OSO_3^-$.

To explain formation of the hydroxy-derivatives as by-products in these reactions, a mechanism has been suggested involving a rearrangement in the intermediate A induced by the catalyst.

$$ArH + SO_4F^- + BF_3 \longrightarrow ArH + OSO_3BF_4^- \longrightarrow ArOH + BF_3 + FSO_3^-$$

The phenol products undergo fluorination or condensation leading to the dimer and polymers. An alternative mechanism of the catalysis, protonation of the SO_4F^- ion, should not be excluded from consideration.

Appelman and his co-workers consider that fluoroaromatic compounds are chiefly formed via intermediate B, whereas the polyfluorinated by-products, and benzyl fluoride (from toluene) are produced by the reactions involving radical cation C. However this does not account for the reason of pronounced *ortho*-fluorination of anisole, toluene, and biphenyl. Even the reaction with fluorobenzene gives a great amount of 1,2-difluorobenzene, though in the electrophilic reactions (halogenation, sulfonation, nitration, etc.) substitution of hydrogen in C_6H_5F proceeds for more than 80% at the *para*-position. Fluorination of benzene derivatives with *meta*-orienting substituents gives about equal or slightly different (NO_2) yields of *ortho*- and *meta*-isomers [41]. In our opinion, this indicates a more complex mechanism of fluorination of aromatic compounds than this is considered by the authors of [41,43]. Thus the isomer ratio here may be affected by the substituent 1,2-shift in the arenonium cation.

Fluorination of alkenylaromatic and related compounds by caesium fluorooxysulfate has been studied. The reactivity of the C=C bond was found to exceed that of the aromatic ring, and $CsSO_4F$ always reacts with the alkenyl fragment, irrespective of its position in a molecule. In the case of 1,1-diphenylethylene, the difluoro-derivative is formed when the reaction is carried out in methylene chloride (yield 74%). In other media, especially in the nucleophilic ones, chiefly the monofluoro-derivatives are formed [46].

$$Ph_2C=CH_2 + CsSO_4F -$$

$$\xrightarrow{CH_2Cl_2, \ 35\ °C} Ph_2C=CF_2 \quad 74\%$$

$$\xrightarrow{HOAc} Ph_2C=CFOAc \quad 48\%$$

$$\xrightarrow{MeOH} Ph_2C=CFOMe \quad 36\%$$

$$\xrightarrow{HF} Ph_2CH-CHF_2 + Ph_2CH-CH_2F \quad 80\% \qquad 20\%$$

This reaction is illustrative of the effect of the nucleophilic solvents involved in it.

It is interesting to note that in α-methylstyrene, the allyl hydrogens are formally substituted, but as a matter of fact the reaction obviously proceeds with isomerisation of the intermediate carbocation.

$$PhMeC=CH_2 + CsSO_4F \rightarrow PhC(CH_2F)=CH_2 + PhC(CHF_2)=CH_2$$
$$ 30\% 32\%$$

In [47], the stereochemical results have been analysed in the "electrophilic" fluorination of acenaphthene, stilbene, indene, and 1-phenylindene by various fluorinating agents, including $CsSO_4F$.

$$55 \quad : \quad 45$$

R = H	59	:	41
Ph	32	:	68

The information on the reactivity of a reagent is usually obtained by carrying out comparative studies on the reactions of such reagents with a model compound. In respect of the fluorinating agents, it would be interesting to compare the reactivity of $CsSO_4F$ with F_2, CF_3OF, CF_3COOF, and XeF_2. As a model, the authors of [42] chose 1,2-diphenylacetylene which had earlier been fluorinated by all these agents.

At a low temperature, the reaction with fluorine in methanol [49] leads to three products: 1,1,2,2-tetrafluoro-1,2-diphenylethane, 1,1,2-trifluoro-2-methoxy-1,2-diphenylethane (main product), and 1,1-difluoro-2,2-dimethoxy-1,2-diphenylethane, whereas CF_3OF at these conditions gives 1,2,2-trifluoro-1-trifluoromethoxy-1,2-diphenylethane as a main product [50]. A similar reaction with CF_3COOF leads to 2-fluoro-1,2-diphenylethanone and dibenzoyl [51], and at room temperature fluorination with XeF_2 (depending on the catalyst used) leads to 1,1,2,2-tetrafluoro-1,2-diphenylethane (in the presence of HF [52]) or to six products (in the presence of trifluoroacetic acid [53]).

The authors of [48] showed that the reaction of 1,2-diphenylacetylene with $CsSO_4F$ in methanol yields two products: 1,1-difluoro-2,2-dimethoxy-1,2-diphenylethane (*11*) and 2,2-difluoro-1,2-diphenylethanone (*12*). In this case compound *12* is formed as a result of sequential transformations of compound *11* in the reaction conditions. This distinguishes $CsSO_4F$ from the above-mentioned fluorinating agents. The authors of [48] studied the regioselectivity of fluorination of other acetylene derivatives by this reagent and found that addition of the methoxy group follows the Markownikov rule for 1-phenylacetylene and 1-phenyl-2-*tert*-butylacetylene, whereas 1-phenyl-1-pentyne leads to

Table 4. Fluorination of phenylacetylenes $PhC \equiv CR$ with $CsSO_4F$ (MeOH, 22 °C) [48]

R	Relative yield (%)							
	11	*12*	*13*	*14*	*15*	*16*	*17*	*11 + 12*
Ph	40	60						100
H	50	50						100
t-Bu	54	46						100
Pr	31	25		25	5	10	4	56
Pr[a]	57	35		8	1	1	1	92
Me	20	19	3	26	6	17	9	39
Me[a]	27	28	2	14	4	16	9	55
Me[b]	41	41	1	12	1	1	5	82

[a] Acetylene : nitrobenzene = 10 : 1, mmol
[b] Acetylene : nitrobenzene = 1 : 1, mmol

1-phenyl-1,1-difluoropentan-2-one (*14*), 1-phenyl-2-fluoropentan-1-one (*15*), 1-phenyl-1-fluoropentan-2-one (*16*), and 1-phenylpentan-1,2-dione (*17*). Similar products are formed from 1-phenyl-1-propyne (together with a small amount of 1-phenyl-1,1-difluoro-2,2-dimethoxypropane (*13*)) (Table 4). The regioselectivity may be explained by the generation of intermediate carbocations, whereas formation of by-products may be attributed to the formation of by-products is confirmed by the experiments of the authors of [48]—they added nitrobenzene and 2,4,6-tri-*tert*-butylphenol as the free-radical traps (Table 4).

$$PhC \equiv CR + CsSO_4F \xrightarrow[22\,°C]{MeOH} PhC(OMe)_2CF_2R + PhC(O)CF_2R$$

	11	*12*
R = H	25%	26%
Ph	23%	32%
t-Bu	23%	19%

Enol acetates are readily transformed by $CsSO_4F$ to α-fluoroketones [52].

There are only few examples of such reactions, among which is successful synthesis of 2α-fluorinated vitamin D_3 [54].

$CsSO_4F$ was used for the synthesis of some fluoro-derivatives of uracyl and uridine [55]. Thus 1,3-dimethyluracyl is transformed to *cis, trans*-5-fluoro-6-methoxy-derivatives to produce 5-fluoro-1,3-dimethyluracyl.

Table 5. Fluorination of aryltrialkylstannanes
$4\text{-}XC_6H_4SnR_3$ with $CsSO_4F$ or fluorine [56]

X	R	Yield (%)	
		$CsSO_4F$	Fluorine
H	Me	69	30
Me	Me	86	57
OMe	Me	79	60
Cl	Me	87	67
OMe	Bu	43	—
Me	Bu	11	—
H	C_6H_{13}	0	47
OMe	C_6H_{13}	0	—
H	Bu	—	41

It is necessary to mention the regioselective fluorodestannylation of aryl-trialkylstannanes to give aryl fluorides in good yields, carried out in [56].

$$4\text{-}XC_6H_4SnR_3 + CsSO_4F \rightarrow 4\text{-}XC_6H_4F$$

The yields of aryl fluorides here are significantly higher than in the reactions with fluorine (Table 5) [56]. With increased size of group at tin, the yield of fluorobenzene decreases, whereas in the reaction with fluorine the case is just the opposite.

The possibilities of caesium fluorooxysulfate are not exhausted in this treatment, and we hope that subsequent investigations will open up its wide potentialities as a fluorinating agent. This agent may be a substrate for the synthesis of monofluoro-derivatives of the aromatic series.

2.4 Acyl Hypofluorites

Fluorination of organic compounds by caesium fluorooxysulfate has analogues in the reactions of acetyl and trifluoroacetyl hypofluorites. The behaviour of organic hypofluorites as fluorinating agents is considered in another chapter, but it would be reasonable to discuss here the results of fluorinations by CH_3COOF and CF_3COOF to compare them with those for $CsSO_4F$. At the same time, these reagents are interesting in their own right. Thus during the last 5 years, acetyl hypofluorite has shown itself to be useful and has been employed for various purposes [57–60]. The need for the samples with radioisotope ^{18}F extended the field of applications of this fluorinating agent [61–66].

Acetyl hypofluorite is prepared by passing elemental fluorine diluted by nitrogen to 5–10%, through a mixture of sodium acetate in anhydrous acetic acid or $CFCl_3$ at $-75\,°C$ [67–69]. It is also synthesized from ammonium [65] or potassium [70] acetates in HOAc at room temperature or by passing fluorine over the solid salt $KOAc·2HOAc$ [71,72]. But fluorination of potassium alkanoates in $CFCl_3$ (or $CF_2ClCFCl_2$, perfluoro-2-butyltetrahydrofuran) leads to a mixture of fluorooxy-compounds. Thus, the potassium perfluorooctanoate yields $CF_3(CF_2)_7OF$, $CF_3(CF_2)_6CF(OF)_2$, and $CF_3(CF_2)_6COOF$. From sodium trifluoroacetate, a mixture of CF_3COOF and CF_3CF_2OF may be obtained [73]. The best results in this reaction (exclusive formation of CF_3COOF) are achieved in HF or H_2O.

In the liquid state, AcOF is an explosive and toxic product, but at $-80\,°C$ and in quantities of less than 0.2 mmol (~ 15 mg) it is quite safe. Physical and spectral data for AcOF are given in [70,74]. If a substrate used for fluorination with AcOF is insoluble in $CHCl_3$, it is possible to use HOAc, $MeNO_2$, and MeCN as solvents [76].

It should be noted that many "electrophilic" fluorination reactions proceed with the formation of by-products via radicals. Therefore the mechanistic conclusions should be made carefully. For example, the reaction of cyclohexane with AcOF in acetic acid gives a mixture of products containing both the products of radical reactions and those of carbocationic reactions. The ratio of these products depends on the reaction conditions [77].

In 1981, Rozen and his co-workers [68] reported fluorination of aromatic compounds with electron-donating substituents by acetyl hypofluorite. Good yields are produced in the fluorination of aryl alkyl ethers, substituted phenols, and acetanilide derivatives, with fluorine substituted for the hydrogen atoms in the *ortho*- and, to a less extent, *para*-position to the electron-donating substituent. Such an unusual isomer ratio is explained by the authors by the addition of AcOF at the $C_{ipso}–C_{ortho}$ bond with subsequent elimination of HOAc.

Likewise, 1-methoxynaphthalene gives a mixture of 1-methoxy-2-fluoro- and 1-methoxy-4-fluoronaphthalenes in the ratio of 6:1 (yield 70%).

The hypothesis about the addition-elimination has been experimentally supported. Thus piperonal (*18*) gives compound *19* which was isolated and its structure was determined by spectral methods.

Benzene, chlorobenzene, toluene, and aniline produce low yields of fluoro-derivatives [78,79] (Table 6).

R= OAlk ,NHCOCH3

The data presented in Table 6 allow one to make two interesting conclusions. First, successful fluorination of an aromatic compound requires the presence of the OAlk or NHCOCH$_3$ substituent. Second, the electronic nature and orienting effect of another substituent in the aromatic ring are unimportant. Thus toluene is transformed to isomeric fluorotoluenes in a total yield of 14%, but 3-trifluoromethylacetanilide gives the products of fluorination at positions 2 and 6, in a 62% yield.

Table 6. Reactions of aromatic compounds with acetyl hypofluorite

Compound	Product	Yield (%)	Ref.
PhOCH$_3$	2-FC$_6$H$_4$OCH$_3$	77	59
	4-FC$_6$H$_4$OCH$_3$	8	
3-CH$_3$OC$_6$H$_4$OCH$_3$	3-CH$_3$O-4-FC$_6$H$_3$OCH$_3$	39	59
	3-CH$_3$O-4,6-F$_2$C$_6$H$_2$OCH$_3$	55	
PhOEt	2-FC$_6$H$_4$OEt	46	59
	4-FC$_6$H$_4$OEt	6	
2-CH$_3$OC$_6$H$_4$NO$_2$	2-CH$_3$O-3,5-F$_2$C$_6$H$_2$NO$_2$	42	59
4-CH$_3$OC$_6$H$_4$NO$_2$	4-CH$_3$O-3-FC$_6$H$_3$NO$_2$	47	59
4-HOC$_6$H$_4$NO$_2$	4-HO-3-FC$_6$H$_3$NO$_2$	62	59

Table 6. (*Continued*)

Compound	Product	Yield (%)	Ref.
2-HOC$_6$H$_4$COOMe	2-HO-3-FC$_6$H$_3$COOMe	9	59
	2-HO-4-FC$_6$H$_3$COOMe	14	
PhNHCOCH$_3$	2-F-C$_6$H$_4$NHCOCH$_3$	55	59
	4-FC$_6$H$_4$NHCOCH$_3$	8	
PhNHCOCF$_3$	2-FC$_6$H$_4$NHCOCF$_3$	57	59
PhNHCOBu-*t*	2-FC$_6$H$_4$NHCOBu-*t*	52	59
2-CF$_3$C$_6$H$_4$NHCOCH$_3$	2-CF$_3$-6-FC$_6$H$_3$NHCOCH$_3$	62	59
3-CH$_3$C$_6$H$_4$NHCOCH$_3$	3-CH$_3$-6-FC$_6$H$_3$NHCOCH$_3$	32	59
	3-CH$_3$-2-FC$_6$H$_3$NHCOCH$_3$	28	
	3-CH$_3$4,6-F$_2$C$_6$H$_2$NHCOCH$_3$	11	
	3-CH$_3$-2,4-F$_2$C$_6$H$_2$NHCOCH$_3$	10	
3-BrC$_6$H$_4$NHCOCH$_3$	2-F-3-BrC$_6$H$_3$NHCOCH$_3$	25	59
	6-F-3-BrC$_6$H$_3$NHCOCH$_3$	47	
3-CF$_3$C$_6$H$_4$NHCOCH$_3$	2-F-3-CF$_3$C$_6$H$_3$NHCOCH$_3$	34	59
	6-F-3-CF$_3$C$_6$H$_3$NHCOCH$_3$	28	
3,5-Me$_2$C$_6$H$_3$NHCOCH$_3$	2-F-3,5-Me$_2$C$_6$H$_2$NHCOCH$_3$	67	59
4-CH$_3$C$_6$H$_4$NHCOCH$_3$	2-F-4-CH$_3$C$_6$H$_3$NHCOCH$_3$	85	59
4-CF$_3$C$_6$H$_4$NHCOCH$_3$	2-F-4-CF$_3$C$_6$H$_3$NHCOCH$_3$	72	59
4-BrC$_6$H$_4$NHCOCH$_3$	2-F-4-BrC$_6$H$_3$NHCOCH$_3$	65	59
3,5-(MeO)$_2$C$_6$H$_3$SnBu$_3$	3,5-(MeO)$_2$C$_6$H$_3$F	68	59
C$_6$H$_5$CH$_3$	2-FC$_6$H$_4$CH$_3$	8	78, 79
	3-FC$_6$H$_4$CH$_3$	1	
	4-FC$_6$H$_4$CH$_3$	4	
	C$_6$H$_5$CH$_2$F	1	
C$_6$H$_5$OH	2-FC$_6$H$_4$OH	45	78, 79
	4-FC$_6$H$_4$OH	30	
C$_6$H$_5$OMe	2-FC$_6$H$_4$OMe	64	78, 79
	4-FC$_6$H$_4$OMe	21	
C$_6$H$_5$NHCOCH$_3$	2-FC$_6$H$_4$NHCOCH$_3$	44	78, 79
	4-FC$_6$H$_4$NHCOCH$_3$	22	
C$_6$H$_6$	C$_6$H$_5$F	18	78, 79
C$_6$H$_5$NH$_2$	2-FC$_6$H$_4$NH$_2$	3.5	78
	4-FC$_6$H$_4$NH$_2$	2.5	
C$_6$H$_5$Cl	2-FC$_6$H$_4$Cl	5	78
	4-FC$_6$H$_4$Cl	5	
C$_6$H$_5$NHCOCH$_3$	2-FC$_6$H$_4$NHCOCH$_3$ $\left.\vphantom{\begin{matrix}a\\b\end{matrix}}\right\}$ 7:1	55	68
	4-FC$_6$H$_4$NHCOCH$_3$		
C$_6$H$_5$OMe	2-FC$_6$H$_4$OMe $\left.\vphantom{\begin{matrix}a\\b\end{matrix}}\right\}$ 9:1	85	68
	4-FC$_6$H$_4$OMe		
3-CH$_3$OC$_6$H$_4$OCH$_3$	2,4-(OMe)$_2$C$_6$H$_3$F $\left.\vphantom{\begin{matrix}a\\b\end{matrix}}\right\}$ 1:2	99	68
	2,4-(OMe)$_2$-1,5-F$_2$C$_6$H$_2$		
4-CH$_3$OC$_6$H$_4$NO$_2$	4-CH$_3$O-3-FC$_6$H$_3$NO$_2$	43	68

This apparently follows from the reaction mechanism which seems to be more complex than suggested by the authors of [59,78].

The field of applications of AcOF as a fluorinating agent for aromatic compounds is rather wide. Though the yields of monofluoroderivatives are not very high, AcOF which is similar in its fluorinating ability to elemental fluorine, seems to exceed it in respect of the mild conditions and selectivity of the fluorinating process.

Apart from the reactions with benzene derivatives, fluorination of methoxynaphthalenes by acetyl hypofluorite has been studied [59].

$$70\%(6:1)$$

2-Methoxynaphthalene gives 65% of 1-fluoro-2-methoxynaphthalene, and 6-methoxyquinoline was converted in a 75% yield to 5-fluoro-6-methoxyquinoline; 6-methoxy-1-tetralone affords 5-fluoro-6-methoxy-1-tetralone (53%) and 5,7-difluoro-6-methoxy-1-tetralone (13%) [59].

Meso-hexaestrol is an important hormone. Its fluorination with AcOF in CHCl$_3$ at $-75\,°C$ affords 40% of the monofluoroderivative and a small amount of the difluoro-derivative.

$$X=Y=H, F \quad X=F \quad Y=H$$

The interaction of acetyl hypofluorite with unsaturated compounds proceeds in mild conditions (-70 to $-80\,°C$) in chloroform or the HOAc–CFCl$_3$ mixture and leads to the fluoroacetoxylation products. Thus cyclohexene reacts with AcOF to form 1-acetoxy-2-fluorocyclohexane (yield 60%), and 1-dodecadecene is transformed to 2-acetoxy-1-fluorododecane (yield 30%) [50,58].

The stereochemistry of addition of acetyl and trifluoroacetyl hypofluorite to olefins has been studied in the case of the reactions with aryl and diarylolefins. The reaction of AcOF with *cis*- and *trans*-stilbenes proceeds stereoselectively, whereas trifluoroacetyl hypofluorite gives the product of stereospecific fluoro-acylation of these olefins (with $R = CF_3$ the reaction gives after hydrolysis the respective hydroxy-compounds).

$$cis\text{-}PhCH=CHPh + RCOOF \rightarrow PhCH(OR)\text{-}CHFPh$$

	erythro : threo (%)	
$R = CH_3$	11	51
CF_3	58	—

$$trans\text{-}PhCH=CHPh + RCOOF \rightarrow PhCH(OR)\text{-}CHFPh$$

	erythro : threo (%)	
$R = CH_3$	7	50
CF_3	—	62

According to the data reported in [73], the presence of electron-accepting groups COMe, COOMe, and Cl in the *para*-position of one of the aromatic rings does not affect the stereoselectivity of the addition of CF_3COOF at a multiple bond. In the case of methylcarbonyl- and methylcarboxyl-substituted stilbenes the addition proceeds regioselectively, whereas the chlorine-containing stilbene forms an equimolar mixture of regioisomers. 2-Carboxymethyl-*trans*-stilbene adds CF_3COOF non-specifically, though the *threo*-conformer is predominant.

$$trans\text{-}RC_6H_4\text{-}CH=CHPh + CF_3COOF \rightarrow RC_6H_4CHF\text{-}CH(OH)Ph$$

$R = 2\text{-COOMe}$	*threo*-40%	*erythro*-14%
4-COOMe	*threo*-80%	
4-COMc	*threo*-28%	
4-Cl*	*threo*-32%	
4-Me*	*threo*-8%	

$$* + \ threo\text{-}4\text{-}RC_6H_4CH(OH)CHFPh \ (R=Cl, \ 32\%; \ R= Me, \ 48\%)$$

It is interesting that treatment of 4-methoxystilbenes both with acetyl and trifluoroacetyl hypofluorites leads to the fluoroacetoxylation products which also contain the fluorine atom in the 3-position.

$$trans\text{-}4\text{-}MeOC_6H_4CH=CHPh \xrightarrow{\ \ CH_3COOF \ or \ CF_3COOF \ \ }$$

$$4\text{-}MeOC_6H_4CH(OAc)CHFPh + 3\text{-}F\text{-}4\text{-}MeOC_6H_4CH(OR)CHFPh$$

erythro-15%	*erythro*-14% (R=Ac)
threo-42%	*threo*-5%
	erythro-57% (R=H, after hydrolysis
	threo-14% of trifluoroacetate)

Treatment of 1,4-androsten-3,17-dione with acetyl hypofluorite leads to the acetoxyfluoro-derivative which eliminates HOAc to give the new fluorodienone steroid [80].

The reaction of AcOF with uracyl and citosine has been investigated [85,86]. In HOAc the reaction gives the fluoroacetoxylation product, whereas in water, hydroxy difluoride is formed. More recently, the fluorination in HOAc with subsequent elimination of the HOAc molecule was used for the synthesis of ^{18}F-labelled uracyl- and citosyl nucleosides [87].

The antipyrine molecule contains the ethylene fragment, and one could expect its fluoroacetoxylation under the action of AcOF. Indeed, this reaction leads to 4-fluoroantipyrine in a 82% yield [80]. In a similar way proceeds fluorination of bimane 24 [88]. If the reaction is carried out in the CHCl$_3$–MeNO$_2$ (2:1) mixture, it gives fluoropyrazolinone 26, along with the mono- and difluorobimanes [76].

The "electrophilic" fluorination of diazepam by AcOF produces 3-fluoro-diazepam whose yield is almost independent of the solvent used. The reaction is supposed to involve the enol form of diazepam [89]. Indeed, the reaction of CF_3COOF with 3-trimethylsiloxydiazepam afforded fluorodiazepam in a 80% yield [90].

Solvent	Yield (%)
HOAc	15
CHCl$_3$	21
CFCl$_3$	18

The ability of acetyl hypofluorite to add at the C=C bond opened up a simple route to fluorinated carbohydrates. Thus, treatment of 3,4,6-tri-O-acetyl-D-glucal with acetyl hypofluorite leads to two isomeric 2-desoxy-2-fluoro-D-glucoses [63,74,91,92].

X=F, OAc, OR$_F$, XeF

Table 8 gives the isomer ratio depending on the fluorinating agent. If AcOF is used in non-polar solvents (CFCl$_3$, CCl$_4$), the yield of products is only 4%, whereas in polar solvents (HOAc, McOH, DMF), the yield is about 20% [93]. The size of the substituent at the hydroxy groups does not affect the yield of product [94]. Detailed analysis of the reaction products revealed compounds that are the typical products of radical reactions [95].

The stereoselectivity of the reaction of AcO ^{18}F with sugars has been studied in [64,95,96].

A remarkable feature of acetyl hypofluorite is the regiospecific fluorode-metallation of organoelement compounds. The most well-studied are the

Table 8. Relative yield of fluorocarbohydrates *28* and *29* obtained from *27* by the action of fluorinating agents [91]

| Fluorinating agent | Solvent | Temperature (°C) | Relative yield (%) | |
			28	*29*
CH_3COOF	$CFCl_3$	−78	95	5
CH_3COOF	HOAc	20	82	18
F_2	$CFCl_3$	−78	80	20
CF_3OF	$CFCl_3$	−78	81	19
$XeF_2–BF_3$	$Et_2O–C_6H_6$	20	93	7

reactions of aromatic derivatives of mercury, tin, silicon (Table 9). The aromatic ring may contain the HO, AlkO, CH_3, F, Br, and OAc substituents, but 3- and 4-fluoroanilines are obtained from the respective arylmercury acetates and AcOF in low yields. The presence of electron-accepting groups (NO_2, COOH) also produces a negative effect on the yields of fluoro-derivatives.

The examples of the reactions with alkylmercury compounds are not numerous [101].

The fluorination of aryltributylstannanes gives higher yields than of aryl-mercuriates. However the latter are more readily available and are therefore better for use in the synthesis of fluoroaromatic compounds.

The solvent produces a significant effect of fluorodemetallation. Thus, in the fluorodestannylation of phenyltrimethylstannane with AcOF, the yield of fluorobenzene at 0 °C is: in $CFCl_3$ 68.2%, in CCl_4 65.5%, in CH_2Cl_2 14.5% [100]. As follows from Table 9, the yield of fluorobenzene decreases in the series: Sn > Ge > Si derivatives. Fluoroaromatic compounds are also obtained in the reaction of acetyl hypofluorite with organogermanium compounds [100], aryltrimethylsilanes [102], and arylpentafluorosilicates [99].

Despite the large volume of information on synthetic aspects of fluorination of organic molecules by acetyl hypofluorite, the data on its chemical behaviuor in various solvents, particularly in acetic acid, are very scarce. Due to this, the mechanistic details of fluorinations by this reagent will have to be cleared up. Most researchers suggest the one-electron oxidation of the aromatic substrate by AcOF in acetic acid [78]. The scheme below shows the possible reaction route. It should be noted that all the products fit into this scheme irrespective of the class of aromatic substrate.

Table 9. Reactions of organoelement compounds with acetyl hypofluorite

Compound	Product	Yield (%)	Ref.
$4\text{-MeOC}_6\text{H}_4\text{HgOAc}$	$4\text{-MeOC}_6\text{H}_4\text{F}$	65	78, 79
$4\text{-MeOCOHgC}_6\text{H}_4\text{NHCOCH}_3$	$4\text{-FC}_6\text{H}_4\text{NHCOCH}_3$	60	78, 79
$2\text{-HOC}_6\text{H}_4\text{HgCl}$	$2\text{-HOC}_6\text{H}_4\text{F}$	53	78, 79
$4\text{-HOC}_6\text{H}_4\text{HgCl}$	$4\text{-HOC}_6\text{H}_4\text{F}$	47	78, 79
$C_6\text{H}_5\text{HgCl}$	$C_6\text{H}_5\text{F}$	55	78, 79
$C_6\text{H}_5\text{HgOAc}$	$C_6\text{H}_5\text{F}$	58	78, 79
$4\text{-NH}_2C_6\text{H}_4\text{HgOAc}$	$4\text{-NH}_2C_6\text{H}_4\text{F}$	4	78, 79
$3\text{-NH}_2C_6\text{H}_4\text{HgOAc}$	$3\text{-NH}_2C_6\text{H}_4\text{F}$	19	78, 79
$4\text{-MeOC}_6\text{H}_4\text{SnBu}_3$	$4\text{-MeOC}_6\text{H}_4\text{F}$	78	97
$4\text{-MeC}_6\text{H}_4\text{SnBu}_3$	$4\text{-MeC}_6\text{H}_4\text{F}$	72	97
$3\text{-MeC}_6\text{H}_4\text{SnBu}_3$	$3\text{-MeC}_6\text{H}_4\text{F}$	71	97
$2\text{-MeC}_6\text{H}_4\text{SnBu}_3$	$2\text{-MeC}_6\text{H}_4\text{F}$	57	97
$C_6\text{H}_5\text{SnBu}_3$	$C_6\text{H}_5\text{F}$	72	97
$4\text{-ClC}_6\text{H}_4\text{SnBu}_3$	$4\text{-ClC}_6\text{H}_4\text{F}$	68	97
$4\text{-FC}_6\text{H}_4\text{SnBu}_3$	$1,4\text{-F}_2C_6\text{H}_4$	73	97
$C_6\text{H}_5\text{SiMe}_3$	$C_6\text{H}_5\text{F}$	10	98
$4\text{-MeC}_6\text{H}_4\text{SiMe}_3$	$4\text{-MeC}_6\text{H}_4\text{F}$	13	98
$4\text{-MeOC}_6\text{H}_4\text{SiMe}_3$	$4\text{-MeOC}_6\text{H}_4\text{F}$	9	98
$4\text{-ClC}_6\text{H}_4\text{SiMe}_3$	$4\text{-ClC}_6\text{H}_4\text{F}$	15	98
$4\text{-BrC}_6\text{H}_4\text{SiMe}_3$	$4\text{-BrC}_6\text{H}_4\text{F}$	14	98
$4\text{-MeCOC}_6\text{H}_4\text{SiMe}_3$	$4\text{-MeCOC}_6\text{H}_4\text{F}$	6	98
$4\text{-AcOC}_6\text{H}_4\text{SiMe}_3$	$4\text{-AcOC}_6\text{H}_4\text{F}$	16	98
$4\text{-Me}_3\text{SiC}_6\text{H}_4\text{SiMe}_3$	$4\text{-FC}_6\text{H}_4\text{SiMe}_3$	16	98
$K_2[C_6\text{H}_5\text{SiF}_5]$	$C_6\text{H}_5\text{F}$	20	99
$K_2[C_6\text{H}_5\text{CH}_2\text{SiF}_5]$	$C_6\text{H}_5\text{CH}_2\text{F}$	6	99
$K_2[4\text{-MeC}_6\text{H}_4\text{SiF}_5]$	$4\text{-MeC}_6\text{H}_4\text{F}$	18	99
$4\text{-MeOC}_6\text{H}_4\text{SnMe}_3$	$4\text{-MeOC}_6\text{H}_4\text{F}$	66	100
$4\text{-MeC}_6\text{H}_4\text{GeMe}_3$	$4\text{-MeC}_6\text{H}_4\text{F}$	16	100
$4\text{-MeC}_6\text{H}_4\text{SiMe}_3$	$4\text{-MeC}_6\text{H}_4\text{F}$	9	100
$C_6\text{H}_5\text{SnMe}_3$	$C_6\text{H}_5\text{F}$	68	100
$C_6\text{H}_5\text{GeMe}_3$	$C_6\text{H}_5\text{F}$	9	100
$C_6\text{H}_5\text{SiMe}_3$	$C_6\text{H}_5\text{F}$	4	100

Acetyl hypofluorite oxidizes the aromatic substrate to form complex D which involves the aryl cation radical, the acetoxyl radical, and the fluoride ion. In the case of organometallic derivative, the metal stabilises the positive charge on the carbon atom and regioselectively gives the fluorodemetallation product. The interaction of the fluoride ion with metal leads to elimination of metal fluoride, and the aryl radical formed is regiospecifically transformed to the aryl acetate, apparently as a result of radical recombination. The acetoxyl radical decomposes to CO_2 and $CH_3^.$ and recombinates to form the toluene derivatives.

Though this route requires further investigations, it allows one to predict the character of products in the fluorinations by acetyl hypofluorite.

A short and simple synthesis of AcOF via elemental fluorine opened up a route to ^{18}F-labelled fluoroorganic compounds. These factors are especially important in view of the half-life period of $^{18}F \sim 110$ min. 3-Methoxy-4-hydroxy-L-phenylalanine reacts with AcO ^{18}F at 20 °C, forming a mixture of 2-, 5-, and 6-fluoroderivatives [103].

Acyl hypofluorites synthesized from perfluorinated carboxylic acids proved to be efficient catalysts and initiators of polymerisation of fluoroolefins [104,105]. Thus polymerisation of tetrafluoroethylene catalysed by perfluoro-octanoyl hypofluorite to give perfluoro-2-butyltetrahydrofurane affords poly(tetrafluoroethylene) in a 89% yield, which is stable up to 372 °C [104].

Thus within a short period of time CH_3COOF became a popular fluorinating agent used inter alia for synthesis of reagents for use in biology. The use of ^{18}F stimulated the development of methods for the synthesis of AcOF and expanded the range of accessible fluoroorganic compounds.

2.5   Conclusion

Choosing a fluorinating agent, a chemist is usually interested in its availability, selectivity of fluorination, equipment for the process, and, of course, the safety of working with it. This review allows one to evaluate these factors, but it seems reasonable to give a general overview of the possibilities of these fluorinating agents.

Among the nitrogen-fluorine-containing compounds the most promising are, apparently, N-fluoropyridinium triflate and its derivatives. These fluorinating agents are relatively easily synthesized, may be stored, and are easy to handle in synthesis. Fluorinations are usually carried out in dichloromethane, dichloro- and trichloroethane at 20 to 100 °C, during several hours (5 to 24 h). This agent may be used to fluorinate aromatic compounds with electron-donating substituents, olefins, enol ethers, and carbanion salts; the processes may be conducted in glassware, as usual. The Japanese researchers [29] conducted fluorination using approx. 1 mmol of N-fluoropyridinium triflate. Other nitrogen-fluorine-containing compounds are either less available or dangerous to handle, or insufficiently studied as fluorinating agents.

Another accessible fluorinating agent is caesium fluorooxysulfate. Zupan and Stavber [42] reported that they synthesized more than 200 g of $CsSO_4F$ (approx. 0.8 g-atom of active fluorine) at a time and stored it in polyethylene flask at $0\,°C$ for 14 days without decrease of its activity. We believe the most remarkable feature of $CsSO_4F$ (or other fluorooxysulfates) to be the predominant *ortho*-fluorination of alkoxyarenes and acetanilides, as the classical methods, such as the Balz-Schiemann reaction, give low yields of such compounds. Using caesium fluorooxysulfate, it is possible to carry out substitution of vinyl hydrogen by fluorine—a rather rare reaction type. It is advisable to use BF_3 as a catalyst. Strong protic acids may cause side reactions, for example, isomerisation.

The use of acyl hypofluorites is certainly justified when it is necessary to synthesize vicinal fluoroacetoxy- (from AcOF) or fluorohydroxy- (from CF_3COOF) containing compounds. α-Fluorocarbonyl compounds are easily obtained in high yields from enol acetates, but fluorination of aromatic compounds with acetyl hypofluorite has no advantages over the use of $CsSO_4F$ or XeF_2. At the same time, the regiospecificity of fluorodemercuration of arylmercury derivatives is an advantage of acetyl hypofluorite, and its use here is preferable. A common disadvantage of CH_3COOF and CF_3COOF is their instability, the necessity of using them immediately after preparation and also restricting the fluorination degree by selecting the amount of acyl hypofluorite (~ 15–100 mg).

The data reviewed allow one to make the following conclusions. First, interest in the development of the method of direct substitution of hydrogen by fluorine in various organic molecules remains as high as ever, and we can expect new results in the future. Second, in addition to the direct fluorination method, new fluorinating agents are being found, which have some advantages over the traditionally used ones. They are widely used for the fluorination of natural and biologically active compounds. Third, the industrial need for new materials with specific features calls for extensive research into such substances among organic compounds. Some fluoroorganic agents have already been found to satisfy these requirements. The development of fluorination methods and search for new fluorinating agents will remain an urgent task in the future.

2.6 Preparations

1. Caesium fluoroxysulfate [42]

In a 100 ml polyethylene vessel containing 10 g of Cs_2SO_4 in 16 ml of water, was introduced a 20% mixture of F_2 in nitrogen at $0\,°C$ for 5 h (complete amount of F_2 introduced was approximately 40 mmol (1.5 g)). After half an hour the insoluble $CsSO_4F$ was filtered off and washed with water (1 ml) and combined precipitates were dried under vacuum at room temperature. Dry $CsSO_4F$ (4 g) was obtained, which must be stored in a polyethylene vessel at $0\,^\circ C$. The compound is stable for at least 14 days, but any contact with a metallic spatula or mechanical pressure must be avoided, since decomposition or even an explosion may take place.

2. Fluorination of acetanilide [42]

$CsSO_4F$ (1 mmol) and 1.5 ml of acetonitrile were stirred at room temperature for 5 min, and after the introduction of BF_3 (0.5–1 mmol) over the reaction mixture, 1 mmol of acetanilide in 0.5 ml of acetonitrile was added. The reaction mixture was stirred at room temperature for 30 min, then 10 ml of dichloromethane was added, the insoluble residue was filtered off, the filtrate washed with water and dried over Na_2SO_4, and the solvent was evaporated under vacuum. The crude reaction mixture (130 mg, 71% conversion of acetanilide) was separated by preparative TLC (SiO_2, dichloromethane : methanol = 9.5 : 0.5). The yield of 2-fluoro-1-(acetylamino)benzene was 80 mg (74.5%), m.p. 76 to 78 °C, and the yield of 4-fluoro-1-(acetylamino)benzene was 12 mg (11%), m.p. 152 to 153 °C.

3. Acetyl hypofluorite in solution [58]

A fluorine-nitrogen mixture ($\sim 10\%$ F_2) was bubbled slowly through a suspension of sodium acetate–acetic acid mixture (14 g; from equimolar amounts of both components) in $CFCl_3$ (450 ml) at $-75\,^\circ C$, which was agitated by an efficient vibromixer. The oxidizing power of the solution was determined by treating aliquots with acidic potassium iodide solution and subsequent titration of the liberated iodine with thiosulfate.

4. ω-Fluoroacetophenone [58]

Diisopropylamine (1.70 ml, 12 mmol) and THF (10 ml) were cooled to $-78\,^\circ C$. Butyllithium as a 15% solution in hexane (7.32 ml, 12 mmol) was added in one portion to produce lithium diisopropylamide and the resultant solution was stirred at $-78\,^\circ C$ for 10 min. Then, a solution of acetophenone (1.202 g, 10 mmol) in THF (2 ml) was added in portions and stirring of the lithium enolate solution was continued at $-78\,^\circ C$ for 15 min and at $25\,^\circ C$ for further 15 min. Diisopropylamine and solvent were then distilled off under reduced pressure at $25\,^\circ C$ and THF (10 ml) was added to the oily residue. The resultant solution was cooled to $-78\,^\circ C$ and added dropwise, quickly, to the stirred acetyl hypofluorite solution (~ 18 mmol) at $-78\,^\circ C$. After 1 min, the mixture was poured into 5% sodium thiosulfate solution (400 ml). The organic layer was washed with conc. $NaHCO_3$ solution (150 ml) and then with water (2×100 ml) until neutral, dried with $MgSO_4$, and evaporated. The crude product was purified by flash vacuum chromatography (silica gel, ethyl acetate : petroleum ether = 1 : 10). The yield of ω-fluoroacetophenone was 0.94 g (68%), m.p. 25 to 26 °C; the recovery of acetophenone was 0.12 g (10%).

5. Fluorination of *meso*-hexestrol dimethyl ether [80]

To the solution of acetyl hypofluorite (15 mmol) was added 2.1 g (7 mmol) of *meso*-hexestrol dimethyl ether dissolved in 40 ml $CHCl_3$. After 10 min the reaction was stopped and worked up as usual. The crude reaction mixture was chromatographed using 20% EtOAc in petroleum ether. Two compounds were obtained and purified by HPLC using 20% EtOAc in cyclohexane. The first one proved to be the mono-fluoro derivative (40% yield), m.p. 140 °C (from EtOAc). The second fraction proved to be the difluoro derivative (20% yield), m.p. 160 °C (from EtOAc).

6. Preparation of N-fluoropyridinium triflates [25]

Method A. The adduct Py·F_2 prepared in $CFCl_3$ at -75°C was allowed to react with sodium triflate in dry acetonitrile at -40 °C for 2 h to give N-fluoropyridinium triflate (6) as very stable white crystals in 67% yield, m.p. 185 to 187 °C.

Method B. It was found that 6 was synthesized conveniently by the reaction of F_2/N_2 (1:9) with pyridine in MeCN at -40 °C followed by treatment with sodium triflate (71%).

Method C. Triflate 7 was prepared by bubbling $F_2/N_2(1:9)$ into a solution of 2,4,6-trimethylpyridine in MeCN at -40 °C in the presence of sodium triflate (49%, m.p. 164 to 166 °C).

7. Fluorination of sulfides (typical procedure) [29]

4-Chlorophenyl methyl sulfide (158 mg, 1 mmol) was added into a mixture of N-fluoro-2,4,6-trimethylpyridinium triflate 7 (290 mg, 1 mmol) and dry dichloromethane (3 ml) at room temperature under argon atmosphere and the reaction mixture was stirred for 8 h. The reaction was followed by checking the oxidation power with aq. KI solution. After triflate 7 completely disappeared, anhydrous sodium carbonate (0.5 g) was added and the organic layer was thin-layer chromatographed on silica gel by using a mixture of hexane and Et_3N (100:1) as an eluent to give 4-chlorophenyl fluoromethyl sulfide as an oily product (133 mg, 76%).

2.7 References

1. Ishikawa N (ed) (1984) Novoe v tekhnologii soedinenii ftora. Russ. per. Mir, Moscow, p 591
2. Ishikawa N, Kobayashi E (1982) Ftor. Khimiya i primenenie. Russ. per. Mir, Moscow, p 91
3. Sheppard WA, Sharts CM (1969) Organic fluorine chemistry. Benjamin, New York
4. Adv. Fluorine Chem. (1961) Butterworth, Washington, vol 2
5. Leidinger S (1984) Bull. Soc. Quim. Peru 50:344
6. Banks RE, Tatlow JC (1986) J. Fluor. Chem. 33:71
7. Riess JC (1987) J. Chem. Phys. 84:1119
8. Haas A, Gerstenberger MRC (1981) Angew. Chem. Int. Ed. Engl. 20:647
9. Haas A, Lieb M (1985) Chimia 39:134
10. Purrington ST, Kagan BS (1986) Chem. Rev. 86:997
11. Zupan M (1984) Vestn. Slov. Kem. Drus. 31:151
12. Cartwright MM, Woolf AA (1984) J. Flour. Chem. 25:263; Christe KO (1984) J. Fluor. Chem. 25:269
13. Yakobson GG, Furin GG (1984) Soviet Sci. Rev. Sect. B. Chem. Rev. 5:255
14. Christe KO, Wilson RD, Goldberg IB (1979) Inorg. Chem. 18:2578
15. Christe KO, Schack CJ, Wilson RD (1976) J. Fluor. Chem. 8:541
16. Schack CJ, Christe KO (1981) J. Fluor. Chem. 18:363

17. Olah GA, Laali K, Farnia M, Shih J, Singh BP, Schack CJ, Christe KO (1985) J. Org. Chem. 50:1338
18. Banks RE, Tsiliopoulos E (1986) J. Fluor. Chem. 34:281
19. Purrington ST, Jones WA (1983) J. Org. Chem. 48:761
20. Purrington ST, Jones WA (1984) J. Fluor. Chem. 26:43
21. Simons JH (1950) In: Simons JH (ed) Fluorine Chemistry. vol 1, Academic, New York, p 420
22. Umemoto T (1988) in: 5th Regular Meeting of Soviet-Japanese Fluorine Chemists, 25–26 Jan 1988, Tokyo, p XI-1
23. Meinert H (1965) Z. Chem. 5:64
24. Meinert H, Cech D (1972) Z. Chem. 12:292
25. Umemoto T, Tomita K (1986) Tetrahedron Lett. 27:3271
26. Umemoto T, Tomita K (1986) Tetrahedron Lett. 27:4465
27. Umemoto T, Onodera K, Tomita K (1986) In: 52nd Nat. Meeting of Japanese Chem. Soc., 1 Apr 1986. Kyoto, abstr. N 1L15
28. Kawada K, Tomita K, Umemoto T (1986) In: 52nd Nat. Meeting of Japanese Chem. Soc., abstr. N 1Z05
29. Umemoto T, Tomizawa G (1986) Bull. Chem. Soc. Jap. 59:3624
30. Banks RE, Boisson R, Tsiliopoulos E (1987) J. Fluor. Chem. 35:13
31. Banks RE, Boisson R, Tsiliopoulos E (1986) J. Fluor. Chem. 32:461
32. Barnette WE (1984) J. Amer. Chem. Soc. 106:452
33. Lee SH, Schwartz J (1986) J. Amer. Chem. Soc. 108:2445
34. US Pat 4479901 (1984); (1985) Chem. Abs. 102:113537
35. Barton DHR, Hasse RH, Pechet MM, Toh HT: J. Chem. Soc. Perkin Trans. I 1974:732
36. US Pat 3917688 (1975); (1976) Chem. Abs. 84:30714
37. Seguin M, Adenis JC, Michand C, Basselier JJ (1980) J. Fluor. Chem. 15:201
38. Singh S, DesMarteau DD, Zuberi SS, Witz M, Huang Hsu-Nan (1987) J. Amer. Chem. Soc. 109:7194
39. Banks RE, Tatlow JC (1986) J. Fluor. Chem. 33:71
40. Appelman EH, Basile LJ, Thompson RC (1979) J. Amer. Chem. Soc. 101:3384
41. Ip DP, Arthur CD, Winans RE, Appelman EH (1981) J. Amer. Chem. Soc. 103:1964
42. Stavber S, Zupan M (1985) J. Org. Chem. 50:3609
43. Appelman EH, Basile LJ, Hayatsu R (1984) Tetrahedron 40:189
44. Stavber S, Zupan M (1981) J. Fluor. Chem. 17:597
45. Visser GWM, Halteren BV, Herscheid JDM, Brinkman G, Hoekstra A (1984) J. Labell. Compounds Radiopharm. 21:1185
46. Stavber S, Zupan M: J. Chem. Soc. Chem. Commun. 1981:795
47. Stavber S, Zupan M (1987) J. Org. Chem. 52:919
48. Stavber S, Zupan M (1987) J. Org. Chem. 52:5022
49. Merrit RF (1987) J. Org. Chem. 32:4124
50. Barton DHR, Danks LJ, Ganguly AK, Hesse RH, Tarzia G, Pechet MM: J. Chem. Soc. Perkin Trans. I 1976:101
51. Stavber S, Zupan M: J. Chem. Soc. Chem. Commun. 1981:148
52. Zupan M, Pollak A (1974) J. Org. Chem. 39:2646
53. Gregorcic A, Zupan M (1979) J. Org. Chem. 44:4120
54. Kobayashi Y, Nakazawa M, Kumadaki I, Taguchi T, Ohshima E, Ikekawa N, Tanaka Y, Deluca HF (1986) Chem. Pharm. Bull. 34:1568
55. Stavber S, Zupan M: J. Chem. Soc. Chem. Commun. 1983:563

56. Bryce MR, Chambers RD, Mullins ST, Parkin A: Bull. Soc. Chim. Fr. 1986:55
57. Lerman O, Rozen S (1983) J. Org. Chem. 48:724
58. Rozen S, Brand M: Synthesis 1985:665
59. Lerman O, Tor Y, Hebel D, Rozen S (1984) J. Org. Chem. 49:806
60. Chirakal R, Firnau G, Gause J, Garnett ES (1984) Int. J. Appl. Radiat. Isotop. 35:651
61. Shine CY, Salvadori PA, Wolf AP (1972) J. Nucl. Med. 23:108
62. Adam MJ: J. Chem. Soc. Chem. Commun. 1983:730
63. Adam MJ, Pate BD, Nesser JR, Hale LD (1983) Carbohyd. Res. 124:215
64. Fowler JS, Shine CY, Wolf AP, Salvadori PA, MacGregor RR (1982) J. Labell. Compounds Radiopharm. 19:1634
65. Shine CY, Salvadori PA, Wolf AP, Fowler JS, MacGregor RR (1982) J. Nucl. Med. 23:899
66. Adam MJ, Ruth TJ, Javan S, Pate BD (1984) J. Fluor. Chem. 25:329
67. Rozen S, Lerman O, Kol M: J. Chem. Soc. Chem. Commun. 1981:443
68. Lerman O, Tor Y, Rozen S (1981) J. Org. Chem. 46:4629
69. Hebel D, Lerman O, Rozen S (1985) J. Fluor. Chem. 30:141
70. Diksic M, Jolly D (1983) Int. J. Appl. Radiat. Isotop. 34:893
71. Jewett DM, Potocki JF, Ehronkaufer RE (1984) Synth. Commun. 14:45
72. Yewett DM, Potocki JF, Ehronkaufer RE (1984) J. Fluor. Chem. 24:477
73. Rozen S, Lerman O (1980) J. Org. Chem. 45:672
74. Appelman EH, Mendelsohn MH, Kim H (1985) J. Amer. Chem. Soc. 107:6515
75. Bida GT, Satyamurthy N, Barrio JR (1984) J. Nucl. Med. 25:1327
76. Kosower EM, Hebel D, Rozen S, Radkowski AE (1985) J. Org. Chem. 50:4152
77. Rozen S, Lerman O, Kol M, Hebel D (1985) J. Org. Chem. 50:4753
78. Visser GWM, Bakker CNM, Halteren BW, Herscheid JDM, Brinkman GA, Hoekstra A (1986) J. Org. Chem. 51:1886
79. Visser GWM, Halteren BW, Herscheid JDM, Brinkman GA, Hoekstra A: J. Chem. Soc. Chem. Commun. 1984:655
80. Hebel D, Lerman O, Rozen S: Bull. Soc. Chim. Fr. 1986:861
81. Barnette WE, Wheland RC, Middleton WJ, Rozen S (1985) J. Org. Chem. 50:3698
82. Rozen S, Menahem Y (1980) J. Fluor. Chem. 16:19
83. Rozen S, Menahem Y: Tetrahedron Lett. 1979:725
84. Shimokawa K (1988) In: 5th Regular Meeting of Soviet-Japanese Fluorine Chemists, 26–27 Jan 1988. Tokyo, p X-1
85. Diksic M, Farrokhzad S, Colebrook LD (1986) Can. J. Chem. 64:424
86. Visser GWM, Boele S, Halteren BW, Knops G, Herscheid JDM, Brinkman GA, Hoekstra A (1986) J. Org. Chem. 51:1466
87. Kosower EM, Hebel D, Rozen S, Radkowcki AE (1985) J. Org. Chem. 50:4152
88. Visser GWM, Noordhuis P, Zwaagstra O, Herscheid JDM, Hoekstra A (1986) Int. J. Appl. Radiat. Isotop. 37:1074
89. Luxen A, Barrio JR, Satyamurthy N, Bida GT, Phelps ME (1987) J. Fluor. Chem. 36:83
90. Middleton WJ, Bingham EM (1980) J. Amer. Chem. Soc. 102:4845
91. Mulholland GK, Ehronkaufer RE (1986) J. Org. Chem. 51:1482
92. Vyplel H (1985) Chimia 39:305
93. Shine CY, Wolf AP (1985) J. Nucl. Med. 26:129
94. Shine CY, Wolf AP (1986) J. Fluor. Chem. 31:255
95. Cornelis JS, Herscheid JDM, Visser GWM, Hoekstra A (1985) Int. J. Appl. Radiat. Isotop. 36:111

96. Adam MJ, Ruth TJ, Javan S, Pate BD (1984) J. Labell. Compounds Radiopharm. 21:11
97. Adam MJ, Ruth TJ, Javan S, Pate BD (1984) J. Fluor. Chem. 25:329
98. Speranza M, Shine CY, Wolf AP, Wilbur DS, Angeloni G (1985) J. Fluor. Chem. 30:97
99. Speranza M, Shine CY, Wolf AP, Wilbur DS, Angeloni G: J. Chem. Soc. Chem. Commun. 1984:1448
100. Coenen HH, Moerlein SM (1987) J. Fluor. Chem. 36:63
101. Habel D, Rozen S (1987) J. Org. Chem. 52:2588
102. Speranza M, Shine CY, Wolf AP, Wilbur DS, Angeloni G (1984) J. Labell. Compounds Radiopharm. 21:1189
103. Chirakal R, Firnau G, Couse J, Garnett ES (1984) Int. J. Appl. Radiat. Isotop. 35:651
104. US Pat 4535136 (1985); (1985) Chem. Abs. 103:178795
105. US Pat 4588796 (1986)

3 Hypofluorites and their Application in Organic Synthesis

Farid Mubarakshevich Mukhametshin

Institute of Applied Chemistry, 614034 Perm, USSR

Contents

3.1 Introduction

Hypofluorites are represented by a wide-spread class of compounds with the O–F functional group bonded to an inorganic or organic residue.

The most well-known representatives of inorganic hypofluorites are oxygen difluoride OF_2, and pentafluorosulfur hypofluorite SF_5OF. Less known are the analogues of the latter—pentafluoroselenium hypofluorite SeF_5OF, pentafluorotellurium hypofluorite TeF_5OF, higher oxygen fluorides such as $FOOF$ and $FOOOF$, and the derivatives of fluorosulfonic FSO_2OF, chloric O_3ClOF, and nitrous O_2NOF acids.

The relative stability of inorganic hypofluorites is explained by the fact that their elements are in the highest oxidation state. At a partial oxidation state, the strong electron-accepting nature of the fluorine atom bonded with oxygen produces a destabilising effect in the element-oxygen-fluorine system. For that reason such compounds as, e.g. F_3SOF and F_2NOF are hypothetical and exist only as the respective oxides $O{=}SF_4$ and $O{\leftarrow}NF_3$.

As opposed to inorganic hypofluorites, the organic ones are much more widely spread, this being again due to their structural peculiarities. Such compounds may be synthesized and may exist in a free form, provided that the OF-bonded carbon substituent is fully or partially (but to a sufficient extent) fluorinated. Their stability is explained by the fluorine screening of the molecular carbon frame, a certain inertness of the C–F bond, and the electron-accepting nature of the perfluoro (or polyfluoro)alkyl substituents. The hydrogen-containing alkyl hypofluorites do not occur in normal conditions. For example, methyl hypofluorite CH_3OF whose formation has been postulated in the reaction of methanol with xenon difluoride [1], was only fixed in situ, as a product of addition to the C=C multiple bond at a low temperature.

The wider occurrence of organic hypofluorites is due to the variety of poly- and perfluorinated compounds. Owing to this, the organic hypofluorites involve not only monofunctional derivatives but also compounds containing two geminal OF fragments, for example difluoromethane-bis-hypofluorite and tetra-fluoroethane-bis-hypofluorite $CF_3CF(OF)_2$. The large capabilities of the per-fluorinated organic compounds are illustrated by several instances of the selective synthesis of hypofluorites with the OF group bonded to the acyl or alkoxy substituents of the type of $CF_3C(O)OF$ and CF_3OOF.

Despite the large number of hypofluorites synthesized up to now, their chemistry is a relatively new field of science, which is not older than 25 to 30 years. Nevertheless, there have been frequent attempts at reviewing the chemistry of hypofluorites, indicating a constant interest in it.

Inorganic hypofluorites were reviewed in [2–8] discussed chiefly inorganic reactions, which are beyond the scope of this review and are not considered here.

Organic hypofluorites were first separated as a subject in the review [9] published in 1980 and devoted to the critical analysis of the current views on the reaction mechanisms and the reactivity towards organic compounds.

This review follows the same trend, developing the previous ideas and covering the literature up till 1987. An attempt is made to explain the available experimental data on the synthesis and chemical properties of organic hypo-fluorites in terms of novel approaches.

3.2   Synthesis of Hypofluorites

The first representative of organic hypofluorites—trifluoromethyl hypofluorite CF_3OF—was synthesized in 1948 by the reaction of fluorine with carbon oxide [10].

$$CO + 2F_2 \rightarrow CF_3OF$$

The reaction was found later to proceed more efficiently in a two-section apparatus with a separate fluorine inlet [11]. One section was intended for the synthesis of carbonyl difluoride: $CO + F_2 \rightarrow COF_2$, in another its further fluorination took place: $COF_2 + F_2 \rightarrow CF_3OF$. The highest temperature for the fluorination of CO is 670 K. Above this temperature, formation of tetra-fluoromethane was reported [12]:

$$CO + 2F_2 \rightarrow CF_4 + 1/2\,O_2$$

Kinetic studies [13] have shown carbon oxide to slowly react with fluorine even at room temperature, the activation barrier being insignificant $(55\,\text{kJ mol}^{-1})$. This is a typical radical–chain reaction, with chain generation proceeding by the bimolecular mechanism: $CO + F_2 \rightarrow \dot{C}OF + F$. Due to its radical-chain nature, the reaction is explosive and hazardous and requires the appropriate safety measures.

It is an important condition for the reaction of $CO + F_2$ that the starting reagents should be oxygen-free. Otherwise, a series of transformations $\dot{C}OF + O_2 \rightarrow \dot{O}OC(O)F \rightarrow \dot{O}C(O)F \rightarrow FC(O)OOC(O)F$ leads to the formation of bis-fluoroformyl peroxide as a by-product, which at elevated temperatures yields such by-products as carbon dioxide, oxygen, and carbonyl difluoride [14].

However, the presence of carbon dioxide in the reaction mixture should not be a constraint, as in an excess of fluorine, CO_2 is known [15] to be fluorinated to trifluoromethyl hypofluorite: $CO_2 + 2F_2 \rightarrow CF_3OF + 1/2\,O_2$. This reaction proceeds smoothly in the temperature range of 473 to 523 K, is explosion-safe unlike fluorination of CO, and therefore may be regarded as a separate method for the preparation of CF_3OF.

It is important to know that fluorination of carbon dioxide as distinct from fluorination of CO requires the preliminary generation of atomic fluorine. The fluorine atoms generated, e.g. photochemically, initially add to the C=O multiple bond, forming carbon- and oxygen-centred radicals, as shown in [16,17].

$$F_2 \xrightarrow{h\nu} 2F; \qquad CO_2 + F \rightarrow O{=}CF\dot{O} + \dot{C}(O)OF$$

The same beginning is expected for the thermal fluorination of CO_2 described in [18]. Thus the mechanism of fluorination of this reagent reminds one of fluorination of carbon oxide in the presence of oxygen.

One of the questions inevitably arising in efforts to rationalise the mechanism of the transformation of CO_2 to CF_3OF is the question of the mechanism of formation of COF_2, which alone can be the source of trifluoromethyl hypofluorite. In the literature this matter has not yet been discussed, but it seems to be solvable in terms of the general approach.

First, only one radical of those formed at the stage of initiation is thermo-dynamically favourable—the oxygen-centred one (the heats of formation vary by about $270\,\text{kJ mol}^{-1}$). Second, it is reasonable to admit that this radical further reacts with fluorine to give fluorooxyformyl fluoride which is converted by atomic fluorine to the fluorooxydifluoromethoxyl radical. The latter undergoes fragmentation to form carbonyl difluoride and the fluoroxyl.

$$O{=}CF\dot{O} + F_2 \rightarrow O{=}CFOF + F \rightarrow \dot{O}CF_2OF \rightarrow O{=}CF_2 + \dot{O}F$$

$$2FO\dot{} \rightarrow O_2 + F_2$$

Such a view of the mechanism of transformation of CO_2 to COF_2 is in agreement with the close reactivities of these reagents towards fluorine. Fluorination of carbonyl difluoride, the best reagent for the synthesis of CF_3OF, proceeds smoothly in the same temperature range as for CO_2. The reaction is an equilibrium, and above 523 K trifluoromethyl hypofluorite slowly decomposes to the starting reagents [19].

When initiated photochemically [20], fluorination of COF_2 is irreversible, but leads to substantial amounts of bis-trifluoromethyl peroxide (up to 20%).

Due to the apparent possibility of the fluorine attack on the carbonyl carbon and oxygen, there are also two viewpoints on the mechanism of fluorination of COF_2. According to one of them, CF_3OF is generated from the oxygen-centred radical $CF_3O^{\cdot}$ [20], according to another—from the carbon-centred radical $^{\cdot}CF_2OF$ [21]. But again, the radical $CF_3O^{\cdot}$ is thermodynamically more favourable.

$$F_2 \xrightarrow{h\nu \text{ or } \Delta} 2F$$

$$F + COF_2 \to CF_3O^{\cdot}$$

$$CF_3O^{\cdot} + F_2 \to CF_3OF + F$$

$$2CF_3O^{\cdot} \to CF_3OOCF_3$$

The intermediate formation of radical $CF_3O^{\cdot}$ is indicated by the above experimental data on fluorination of carbon oxide by fluorine excess where the reaction produces tetrafluoromethane in accordance with the scheme.

$$CF_3O^{\cdot} + CO \to CF_3OC^{\cdot}(O) \to {}^{\cdot}CF_3 + CO_2$$

$$^{\cdot}CF_3 + F_2 \to CF_4 + F$$

The latter scheme was confirmed by studies of the thermal reaction of bis-trifluoromethyl peroxide and carbon oxide, where the main products are carbon dioxide and hexafluoroethane [22].

$$CF_3OOCF_3 + CO \to CO_2 + C_2F_6$$

It should be noted that the above methods of thermal and photochemical generation of fluorine may only serve to obtain CF_3OF. The higher organic hypofluorites may not be obtained by this method, as the starting compounds are converted by fluorine at high temperature or under irradiation to the respective perfluoroalkanes (see, e.g. [23]). The process may be directed as desired by using certain catalysts.

The catalytic method plays a leading role in the synthesis of perfluoroalkyl hypofluorites. The recommended heterogeneous catalysts are the sodium, potassium, rubidium, caesium, magnesium, calcium, strontium, barium, brass, iron, copper, nickel, silver, and cobalt fluorides [24]. The homogeneous catalysts are represented by nitrosyl fluoride [25].

The highest catalytic activity in the above series of fluorides is shown by caesium fluoride. It is effective for the whole series of perfluorinated carbonyl-containing reagents (excluding carbon oxide).

Thus perfluoroacyl fluorides and perfluoroketones in the presence of CsF give the primary and secondary perfluoroalkyl hypofluorites respectively [26–32].

$$R_FCF_2C(O)F + F_2 \xrightarrow{CsF} R_FCF_2CF_2OF$$

$$(R_F)_2C{=}O + F_2 \xrightarrow{CsF} (R_F)_2CFOF$$

The extremely high catalytic activity and selectivity of CsF make it effective even with such unstable compounds as peroxides [31,32].

$$FC(O)OOC(O)F + F_2 \xrightarrow{CsF} FOCF_2OOCF_2OF$$

$$CF_3OOC(O)F + F_2 \xrightarrow{CsF} CF_3OOCF_2OF$$

Another example is the synthesis of difluoromethane-bis-hypofluorite with a high yield, by fluorination of carbon dioxide in mild conditions, which has only become feasible due to the high catalytic activity of CsF [27,34–36].

$$CO_2 + 2F_2 \xrightarrow{CsF} CF_2(OF)_2$$

The preparative syntheses of hypofluorites with the use of caesium fluoride show nearly quantitative yields of target products. But to achieve the maximal efficiency, the catalyst should be carefully dried. This procedure is generally carried out in the reactor, by heating the catalyst to 500 to 550 K, with simultaneous evacuation of the reactor for many hours. Fluorine and the carbonyl-containing reagent are then mixed in the reactor at 77 K and are allowed to slowly warm to room temperature. Strict dosing of the reagents removes the need for additional purification of the products.

The catalytic action of CsF is generally recognised to be based on its ability to form alkoxides with perfluoroacyl fluorides [37,38].

$$R_FC(O)F + CsF \rightleftarrows R_FCF_2O^- Cs^+$$

Until recently, the transformation of the alkoxides to hypofluorites had been considered (see, e.g. [39]) to result from the nucleophilic S_N2 substitution of fluorine in a molecule by the alkoxide.

$$R_FCF_2O^- + F_2 \rightarrow R_FCF_2OF + F^-$$

This opinion is now considered erroneous. As shown in [40], in reality the system $COF_2 + F_2 + CsF$ generates the trifluoromethoxyl radicals, which may be recorded by adding carbon oxide into the reaction mixture to obtain tetrafluoromethane and carbon dioxide in accordance with the above scheme. The generation of the $CF_3O^{\cdot}$ radicals is considered in [40] to be best explained by the one-electron transition.

$$CF_3\overset{\frown}{O}Cs^+ + F-F \xrightarrow{-e} [CF_3O^{\cdot} \ldots Cs^+ \ldots F^- \ldots F^{\cdot}] \rightarrow CF_3O^{\cdot} + CsF + F$$

The radical mechanism of the reaction of fluorine with anionoid species is also indicated by the results of fluorination of carbon dioxide in the presence of CsF. This reaction almost always produces insignificant amounts of trifluoromethyl hypofluorite along with difluoromethane-bis-hypofluorite, and may only be rationalised on the assumption that the system generates carbonyl difluoride.

The initial stage of the catalytic fluorination of CO_2 is formation of the alkoxide, which has been strictly proved [41].

$$O=C=O + CsF \rightleftarrows O=CFO^- Cs^+$$

Subsequent interaction of the alkoxide and fluorine by the one-electron transition mechanism leads to generation of the oxygen-centred radical in the system.

$$O=CFO^- Cs^+ + F_2 \xrightarrow{-e} O=CFO^{\cdot} + CsF + F$$

Further on this radical is transformed, by the same route as in the thermal fluorination of CO_2, to the fluorooxydifluoromethoxyl radical, which either reacts with fluorine or undergoes minor fragmentation, giving carbonyl difluoride. Apart from the gas-phase route of the reaction, a considerable role is presumably played by the heterophase route which involves the intermediate formation of a complex of CsF with fluorooxyformyl fluoride.

$$O=CFO^{\cdot} + F_2 \rightarrow O=CFOF + F$$

$$O=CFOF + CsF \rightleftarrows Cs^{+\ -}OCF_2OF$$

$$Cs^{+\ -}OCF_2OF + F_2 \rightarrow {^{\cdot}}OCF_2OF + CsF + F$$

$$^{\cdot}OCF_2OF \begin{array}{c} \longrightarrow CF_2O + FO^{\cdot} \\ \underset{F_2}{\boxed{}} \\ \longrightarrow FOCF_2OF + F \end{array}$$

A shift of the reaction towards synthesis of difluoromethane-bis-hypofluorite is achieved by conducting it at a low temperature (heating of the mixture of the reagents from 77 to 195 K).

In comparison to caesium fluoride (and, possibly, potassium fluoride), the catalysis by other metal fluorides is not so effective. In any case, the literature contains no data on the use of other heterogeneous catalysts in the synthesis of higher perfluoroalkyl hypofluorites, and their activity was estimated in the case of fluorination of carbonyl difluoride.

The fluorides of alkaline earth and transition metals are essentially much less active than caesium fluoride, and fluorinations in their presence require temperatures from 300 to 400 K [24]. The fluorides of transition metals show a greater activity than those of alkaline earth metals.

The catalytic activity of transition metal fluorides is directed to a different object of catalysis. As opposed to CsF activating the carbonyl-containing reagent, they activate fluorine. According to [42], the role of a catalyst consists in atomisation of fluorine, with the subsequent radical (but not chain) process described by the scheme.

$$F_2 \xrightarrow{Ct} 2F_{gas}$$

$$F_{gas} + COF_2 \rightarrow CF_3O^.$$

$$CF_3O^. + F_{gas} \rightarrow CF_3OF$$

$$2CF_3O^. \rightarrow CF_3OOCF_3$$

The principal arguments in support of this mechanism are the observed inability of transition metal fluorides to form alkoxides and the presence of bis-trifluoromethyl peroxide among the reaction products. Nevertheless there was another viewpoint suggesting formation of an intermediate complex from, e.g. silver difluoride, with participation of carbonyl difluoride [43].

$$AgF_2 + COF_2 \rightarrow Ag(OCF_3)_2 \xrightarrow{F_2} [Ag(OCF_3)_2F_2]$$

$$\rightarrow AgF_2 + CF_3OOCF_3$$

The existence, for more than 15 years, of two conflicting opinions on the mechanism of catalysis by metal fluorides seems to be an obvious flaw in elemental fluorine chemistry, which stimulated the author of this review to present his own view, which seems to be well substantiated by the following facts.

The fluorides of transition metals, such as cobalt, silver, and nickel are known to react readily with fluorine, forming higher fluorides: CoF_3 [44], AgF_2 [44], and $NiF_{2.22}$ [45]. To cause the reverse reaction of fluorine elimination, the above higher fluorides should be heated to a temperature at least as high as 500 K. But fluorination of carbonyl difluoride on a catalyst, e.g. AgF_2 [43], proceeds at a temperature 100 K lower than that. It means that the catalyst does not generate the atomic fluorine into the gas phase, but only chemosorbs it, thus causing the activation (weakening or cleavage) of the F–F bond. In the case of NiF_2, the chemosorption of fluorine has been revealed by special experiments, and starts at the temperature of > 373 K. As for silver difluoride, it can fluorinate, at 373 K, not only carbonyl difluoride but also carbon oxide.

In the light of the above facts, it is reasonable to suggest that the radical species are formed in the system "transition metal fluoride–fluorine–reagent" as a result of the interaction of the reagent with the centre on the catalyst surface, at which fluorine is sorbed, i.e. by a two-step mechanism.

$$F_2 \xrightarrow{Ct} 2F_{adsorb}$$

$$F_{adsorb} + COF_{2,gas} \rightarrow CF_3O^._{gas}$$

Because of by-products and difficulties with subsequent separation and purification of the products, the above fluorination may hardly be recommended for the synthesis of individual hypofluorites. At the same time, they must not be completely ignored. Some of these methods (in particular, fluorination of trifluoroacetates at a decreased temperature) use the fluorinating agents that are milder and more selective than fluorine, and which may fluorinate organic compounds in suitable solvents. Another example of the rejection of the direct fluorination method is the transformation of sodium perfluoro-*tert*-butoxide, giving a high yield [50].

$$(CF_3)_3CONa + F_2 \rightarrow (CF_3)_3COF + NaF$$

The similarity of chemical behaviour, the absence of critical discrepancies between the procedures, and theoretical considerations necessitate discussion in this review of one more interesting class of the derivatives containing the OF group—fluoroperoxides.

The only convenient preparative method for the synthesis of fluoroperoxides is considered to be the caesium fluoride-catalysed reaction, where one of the reagents is a carbonyl-containing compound, another—oxygen difluoride. Thus, with the use of COF_2 as carbonyl containing reagent, trifluoromethyl fluoroperoxide is formed with a high yield [61].

$$COF_2 + OF_2 \xrightarrow{\text{CsF}} CF_3OOF$$

Theoretically, it is essential that the reaction be considered in terms of the nucleophilic S_N2 substitution of fluorine in the oxygen difluoride molecule, as in the synthesis of hypofluorites [62]. This is supported by serious arguments. In particular, it has been shown that in the reaction of ^{17}O-labelled carbonyl difluoride with OF_2, the isotope label of the peroxide formed in the reaction is part of the trifluoromethoxyl fragment of the molecule.

$$C^{17}OF_2 + OF_2 + \xrightarrow{\text{CsF}} CF_3{}^{17}O\text{--}OF$$

On the other hand, when $^{17}OF_2$ and COF_2 are made to react, the product is the peroxide with the isotope label in the fluorooxy fragment.

$$COF_2 + {}^{17}OF_2 \xrightarrow{\text{CsF}} CF_3O\text{--}{}^{17}OF$$

Though the arguments seem to be attractive, the question should be considered open owing to the possibility of the alternative interpretation of the reaction in terms of the one-electron transfer, as may be seen from the following schemes.

$$C^{17}OF_2 + CsF \rightleftarrows Cs^+\,{}^-({}^{17}O)CF_3 \xrightarrow[-e,\,-CsF]{FOF} {}^{\cdot}({}^{17}O)CF_3 + {}^{\cdot}OF$$

$$^{\cdot}({}^{17}O)CF_3 + {}^{\cdot}OF \rightarrow CF_3{}^{17}O\text{--}OF$$

or

$$COF_2 + CsF \rightleftarrows CF_3O^-Cs^+ \xrightarrow[-e,\,-CsF]{F^{17}OF} CF_3O^{\cdot} + {}^{\cdot}({}^{17}O)F$$

$$CF_3O^{\cdot} + {}^{\cdot}({}^{17}O)F \rightleftarrows CF_3O{-}^{17}OF$$

Both schemes reproduce the peculiarities of the catalytic synthesis of trifluoromethyl hypofluorite and difluoromethane-bis-hypofluorite from COF_2 and CO_2 respectively. The necessary (and principal) condition for these reactions is the low reaction temperature.

Quite serious arguments in favour of the mechanism involving the fluoro-oxide radicals are suggested [63,64]. The former shows that oxygen difluoride reacts with trifluoromethane to form trifluoromethyl hypofluorite.

$$CHF_3 + OF_2 \rightarrow CF_3OF$$

The latter studies the reaction of O_2F_2 with perfluoropropylene giving the following fluoroperoxides.

$$CF_3CF{=}CF_2 + FOOF \rightarrow CF_3CF_2CF_2OOF + CF_3CF(OOF)CF_3$$

Participation of fluorooxide radicals in the reaction is also indicated by the results of an investigation [65], where the UV irradiation of a mixture of oxygen, difluoride with dichlorotetrafluoroacetone led to chlorodifluoromethyl hypofluorite.

$$(ClCF_2)_2CO \rightarrow ClCF_2C^{\cdot}O + ClF_2C^{\cdot}$$
$$\underset{-CO}{\underbrace{\qquad\qquad}}$$

$$ClF_2C^{\cdot} + FOF \rightarrow ClCF_3 + {}^{\cdot}OF$$

$$ClF_2C^{\cdot} + {}^{\cdot}OF \rightarrow ClCF_2OF$$

The one-electron transfer mechanism is also possible for the fluorination of perfluorocarboxylic acids in the presence of CsF leading to bis-hypofluorites [66].

$$R_FCOOH + F_2 \xrightarrow{CsF} R_FCF(OF)_2$$

$$R_F = CF_3,\ C_2F_5,\ C_3F_7$$

3.3 Physical Properties and Structure

The elementary knowledge on the nature of fluoroorganic hypofluorites is provided by the acquaintance with the physical properties and structure of the most fully described and simplest compounds–trifluoromethyl hypofluorite and difluoromethane-bis-hypofluorite.

Trifluoromethyl hypofluorite is a gas (b.p. 178 K). Liquid CF_3OF has a density of $1.9\ \mathrm{g\,cm^{-3}}$ and solidifies only at 63 K or below. The standard heat of rate. Decomposition products depend on the reactor material and heating conditions.

If CF_3OF is heated in a nickel reactor pre-passivated by fluorine, reversible decomposition occurs [94].

$$CF_3OF \rightleftharpoons COF_2 + F_2'$$

The reaction is accompanied by the formation of considerable amounts of bis-trifluoromethyl peroxide, indicating the initial formation of the trifluoromethoxyl radicals. Theoretical treatment of this question is presented in works [95,96].

At temperatures above 673 K, decomposition of CF_3OF becomes irreversible due to the reaction of the liberated fluorine with the reactor walls. With the carbonyl difluoride additions, decomposition of CF_3OF proceeds more slowly.

The effect of glow discharge on trifluoromethyl hypofluorite differs in its results from ordinary heating: in a quartz reactor this gives CF_4, CO_2, O_2, and SiF_4; in a metal one—CF_4, CO_2, COF_2, and F_2 [94].

As opposed to CF_3OF, thermolysis of difluoromethane-bis-hypofluorite is irreversible under any conditions. Slow heating of $CF_2(OF)_2$ leads to the formation of the products similar to the products of decomposition of CF_3OF.

$$CF_2(OF)_2 \xrightarrow{\Delta} COF_2 + F_2 + 0.5\,O_2$$

This result is completely in agreement with the above-suggested mechanism of fluorination of carbon dioxide with participation of the fluorooxydifluoromethoxyl radical. Only here the radical is formed as a result of decomposition of the bis-hypofluorite at the first stage of the reaction.

$$CF_2(OF)_2 \xrightarrow{\Delta} FOCF_2O^{\cdot} + F$$

In conditions of explosive decomposition of $CF_2(OF)_2$ [77], radical $FOCF_2O^{\cdot}$ undergoes more complex transformations.

$$CF_2(OF)_2 \xrightarrow{explosion} COF_2 + CF_3OF + CF_3OOCF_3 + CF_3OOF + CF_4$$

A distinction of thermal decomposition of higher perfluoroalkyl hypofluorites is the chain character of the reaction. It raises the tendency to stabilisation of the alkoxy radicals by the homolytic decomposition of the C–C bond with subsequent formation of alkanes [49].

$$CF_3CF_2OF \xrightarrow{\Delta} CF_3CF_2O^{\cdot} + F$$

$$CF_3CF_2O^{\cdot} \rightarrow CF_3^{\cdot} + COF_2$$

$$^{\cdot}CF_3 + CF_3CF_2OF \rightarrow CF_4 + CF_3CF_2O^{\cdot} \quad \text{etc.}$$

In accordance with this tendency, secondary and tertiary hypofluorites are transformed to acyl fluorides and ketones.

$$(CF_3)_2CFOF \xrightarrow{\Delta} (CF_3)_2CFO^{\cdot} + F$$

$$(CF_3)_2CFO^{\cdot} \rightarrow CF_3C(O)F + \dot{C}F_3$$

$$(CF_3)_3COF \xrightarrow{\Delta} (CF_3)_3CO^{\cdot} + F$$

$$(CF_3)_3CO^{\cdot} \rightarrow (CF_3)_2CO + {}^{\cdot}CF_3$$

Thus higher perfluoroalkyl hypofluorites, as well as difluoromethane-bis-hypofluorite, are less thermally stable than CF_3OF. There is a linear correlation between the O–F bond energy and the ^{19}F chemical shift [96]. Due to this, for the tentative estimation of thermal stability of hypofluorites we may use the NMR spectral data.

3.4.2 Photochemical Decomposition

The synthetic purpose in the photolysis of hypofluorites is to generate the perfluoroalkoxyl radicals in mild conditions. As the O–F bond cleavage energy is small, any UV source may be used for this purpose. Though very short waves may produce an undesirable effect. For example, photolysis of difluoromethane-bis-hypofluorite at $\lambda = 254$ nm (~ 460 kJ mol^{-1}) results in cleavage of the O–F and C–OF bonds [97], whereas at long waves the process follows its usual route [98].

$$CF_2(OF)_2 \xrightarrow{hv} COF_2 + F_2 + 0.5\,O_2$$

The photolysis of CF_3OF in gas phase yields bis-trifluoromethyl peroxide [20]. In the argon matrix (8 K, $\lambda < 400$ nm), the photolysis is accompanied by the formation of COF_2. The ESR control of irradiation of the mixtures of CF_3OF with NF_3 or CF_4 at 77 K allowed the detection of radical $CF_3OO^{\cdot}$ [98].

On the basis of the authors' suggestion about the presence of an oxygen trace in the mixture, it is possible to admit that the trifluoromethoxyl radical is first converted to the trioxide one, which subsequently recombinates.

$$CF_3OF \xrightarrow{hv} CF_3O^{\cdot} + F$$

$$CF_3O^{\cdot} + O_2 \rightarrow CF_3OOO^{\cdot} \rightarrow CF_3OO^{\cdot} + 0.5\,O_2$$

This route of the reaction is supported, first, by the studies of the photo-sensibilised oxidation of CO in CO_2 by the UV irradiation of a mixture of $CF_3OF + CO + O_2$, where the quantum yield of carbon dioxide reaches

$\sim 10^4$ molecules $(h\nu)^{-1}$ [99], and, second, by the results of [100].

$$CF_3O^{\cdot} + O_2 \rightarrow CF_3OOO^{\cdot}$$

$$CF_3OOO^{\cdot} + CO \rightarrow CF_3OO^{\cdot} + CO_2$$

$$CF_3OO^{\cdot} + CO \rightarrow CF_3O^{\cdot} + CO_2$$

$$CF_3O^{\cdot} + CO \rightarrow {}^{\cdot}CF_3 + CO_2$$

$${}^{\cdot}CF_3 + O_2 \rightarrow CF_3OO^{\cdot} \quad \text{etc.}$$

3.4.3 Reactions with Carbonyl-Containing Compounds

Interaction of hypofluorites with carbonyl-containing compounds is a method for the synthesis of alkyl peroxides, ethers and esters, carbonates, etc. It has been most fully studied in the case of the reaction of CF_3OF and $CF_2(OF)_2$.

Most known reactions of this type require the thermal or photochemical pre-activation of the O–F bond. Besides, some cases of activation by catalysts have been reported, and some of the reactions proceed on their own.

In particular, thermal or photochemical initiation is required for the reaction of CF_3OF with COF_2, leading to the formation of bis-trifluoromethyl peroxide [101–110].

$$CF_3OF + COF_2 \xrightarrow{\Delta} CF_3OOCF_3$$

In conditions of thermal initiation the process is reversible. The optimal temperature for the synthesis of the peroxide both in the static [102,103] and flow [43] conditions is in the range of 500 to 550 K. Up to 520 K the equilibrium is shifted to the right. The system $CF_3OF + COF_2$ obeys the Le Chatelier principle: at the pressure of 10 MPa and the temperature of 550 K, yield of the peroxide reaches 91% [102]. As shown by the kinetic measurements [105–107], the limiting stage of the thermal synthesis of CF_3OOCF_3 is the decomposition of trifluoromethyl hypofluorite.

$$CF_3OF \xrightarrow{\text{slow}} CF_3O^{\cdot} + F$$

$$COF_2 + F \xrightarrow{\text{fast}} CF_3O^{\cdot}$$

$$2CF_3O^{\cdot} \xrightarrow{\text{fast}} CF_3OOCF_3$$

With UV irradiation, the process follows the same route.

The photodecomposition of CF_3OF ($\lambda = 360$ nm) in the presence of carbon oxide leads to the formation of trifluoromethyl fluoroformiate and trifluoromethyl oxalate [111–113].

$$CF_3O^{\cdot} + CO \rightarrow CF_3O\dot{C}O$$

$$CF_3O\dot{C}O + CF_3OF \rightarrow CF_3OC(O)F + CF_3O^{\cdot}$$

$$2CF_3O\dot{C}O \rightarrow CF_3OC(O)C(O)OCF_3$$

For the analogous thermal reaction at 353–383 K, a mechanism of self-initiation of chains has been suggested on the basis of the fact that in these conditions CF_3OF does not tend to dissociate [114].

$$CF_3OF + CO \xrightarrow{\Delta} CF_3O^{\cdot} + {}^{\cdot}COF$$

A convincing proof of the bimolecular initiation is the low activation energy of the total process (75 kJ mol^{-1}). If CF_3OF decomposed at the O–F bond, the activation energy would be at least as high as the energy of the bond cleavage, i.e. it would be more than twice as high as it really is.

Another route in the reaction of trifluoromethyl hypofluorite with carbon oxide is the synthesis of bis-trifluoromethyl carbonate.

$$CF_3O^{\cdot} + CF_3O\dot{C}O \rightarrow (CF_3O)_2CO$$

The latter may also be obtained in a small yield by the photolysis of a mixture of CF_3OF with hexafluoroacetone [115].

The reaction of CF_3OF with carbonyl-containing compounds indicates both an extremely high reactivity of the trifluoromethoxyl radical and its relative stability, which is not the case with the alkoxyl radical formed from $CF_2(OF)_2$. Under conditions of thermal initiation the reaction of $CF_2(OF)_2$ with CO gives carbon dioxide and carbonyl difluoride in the $1:2$ mole ratio [116]. The activation energy value (83 kJ mol^{-1}) and the reaction temperature (383–400 K), which are similar to those for CF_3OF, indicate the same chain initiation stage for both processes [116].

$$CF_2(OF)_2 + CO \rightarrow FOCF_2O^{\cdot} + {}^{\cdot}COF$$

Thus, here again we come across the $FOCF_2O^{\cdot}$ radical, whose transformation to CO_2 and COF_2 for the system $CF_2(OF)_2 + CO$ seems to be quite legal from the point of view of the above-considered schemes.

$$CF_2(OF)_2 + {}^{\cdot}COF \rightarrow FOCF_2O^{\cdot} + COF_2$$

$$FOCF_2O^{\cdot} \rightarrow {}^{\cdot}OF + COF_2$$

$$FO^{\cdot} + CO \rightarrow FO\dot{C}O \rightleftarrows CO_2 + F$$

Proceeding to draw the analogy between the catalytic systems of hypofluorites and their catalysed reactions with carbonyl-containing reagents, it should be noted that the latter reactions also have the radical nature. The main products here are also the alkyl peroxides; the catalytic effect is produced by the silver, copper, nickel, mercury, cobalt, iron [103], and caesium fluorides [43].

The catalytic effect of these fluorides is obviously based on the generation of the $FOCF_2O^{\cdot}$ radicals as a result of chemosorption of atomic fluorine and subsequent fluorination of the carbonyl-containing reagent. Thus synthesis of bis-trifluoromethyl peroxide from CF_3OF and COF_2, in the presence of silver fluoride may be represented by the following scheme.

$$CF_3OF_{gas} + AgF \rightarrow CF_3O'_{gas} + AgF_2(F_{ads})$$

$$COF_2 + AgF_2 \rightarrow CF_3O'_{gas}$$

$$2CF_3O'_{gas} \rightarrow CF_3OOCF_{3\,gas}$$

Consequently, like the similar synthesis of CF_3OF, the total process of the catalytic synthesis of CF_3OOCF_3 is the heterogeneous process.

As shown in [40], for the CsF-catalysed processes, the mechanism of formation of both CF_3OF and CF_3OOCF_3 is again the same. Synthesis of CF_3OOCF_3 here is also preceded by the generation of the CF_3O' radical, which may be detected by adding carbon oxide to the reaction mixture to produce CO_2 and CF_4 (see above).

The most satisfactory explanation of this result is offered by the one-electron transfer theory. The role of electron donor here is played by caesium trifluoromethoxide, that of acceptor—by trifluoromethyl hypofluorite. The electron transfer proceeds with the formation of two CF_3O' radicals, which subsequently recombine [40].

$$CF_3\ddot{O}^- Cs^+ + \overset{\frown}{F}-OCF_3 \xrightarrow{-e} [CF_3\dot{O}^\uparrow \cdots Cs^+ \cdots F^- \cdots \dot{O}CF_3] \longrightarrow CF_3\dot{O}^\uparrow_{gas} + CsF + CF_3\dot{O}^\downarrow_{gas}$$

It is interesting that in the transition state free electrons of the CF_3O fragments are in the fields of opposite charge. Owing to this, the electron spins are opposite, and recombination is possible.

It should be noted that the one-electron transfer theory offers a simple explanation of some other experimental facts as well, which seem anomalous at first sight. This concerns, first of all, the reaction of $CF_2(OF)_2$ with COF_2 in the presence of CsF. From the viewpoint of the S_N2 mechanism, the main product in this reaction should be expected to be CF_3OOCF_3. As a matter of fact, the main products are bis-trifluoromethyl trioxide and trifluoromethylfluoroformyl peroxide [117–119].

$$CF_2(OF)_2 + COF_2 \xrightarrow{CsF} \begin{cases} X \longrightarrow CF_2(OOCF_3)_2 \\ \longrightarrow CF_3OOOCF_3 + CF_3OOC(O)F \end{cases}$$

The one-electron transfer accounts for the result observed in the following series of radical transformations.

$$FOCF_2OF + CF_3O^-Cs^+ \xrightarrow{-e} FOCF_2O' + CsF + CF_3O'$$

$$FOCF_2O' \longrightarrow COF_2 + 'OF \xrightarrow{'OF} F + F + O_2$$

$$F + FOC(O)F \xrightarrow{COF_2} CF_3O' + 'OC(O)F \longrightarrow CF_3OOC(O)F$$

$$CF_3O' + O_2 \longrightarrow CF_3OOO' \longrightarrow CF_3OO' + 0.5O_2$$

$$CF_3O' + CF_3OO' \longrightarrow CF_3OOOCF_3$$

A similar scheme may be presented to explain the formation of the products of the reaction of potassium perfluoro-*tert*-butoxide with trifluoromethyl hypofluorite [120].

$$(CF_3)_3CO^-K^+ + CF_3OF \xrightarrow[-e]{} (CF_3)_3CO^{\cdot} + KF + CF_3O^{\cdot}$$

$$CF_3O^{\cdot} \rightarrow F + COF_2 \xrightarrow{KF} CF_3O^-K^+$$

$$(CF_3)_3CO^-K^+ + COF_2 \rightarrow (CF_3)_3COC(O)F \xrightarrow{(CF_3)_3CO^-K^+} [(CF_3)_3CO]_2CO$$

$$(CF_3)_3CO^{\cdot} + F \rightarrow (CF_3)_3COF$$

Lastly, it is necessary to mention two other "anomalous" results. One case is the reaction of CF_3OF with difluorocarbamyl fluoride, where, instead of the expected peroxide, the reaction product is *O*-trifluoromethyl-*N,N*-difluoro-hydroxylamine [121].

$$F_2NC(O)F + CF_3OF \xrightarrow{CsF} \begin{cases} X \longrightarrow F_2NCF_2OOCF_3 \\ \longrightarrow F_2NOCF_3 \end{cases}$$

Another case is the intramolecular cyclization of perfluoroglutaryl difluoride by fluorine [122].

$$FC(O)(CF_2)_3C(O)F + F_2 \xrightarrow{CsF} O\begin{array}{c} \diagup CF_2-CF_2 \\ | \\ \diagdown CF_2-CF_2 \end{array}$$

It is impossible to explain these results in terms of the ionic mechanism, whereas one-electron transfer gives a quite convincing explanation of them.

$$F_2NC(O)F + CsF \longrightarrow F_2NCF_2O^-Cs^+ \xrightarrow[-e]{CF_3OF} F_2NCF_2O^{\cdot} + CsF + CF_3O^{\cdot}$$

$$F_2NCF_2O^{\cdot} \longrightarrow COF_2 + {}^{\cdot}NF_2 \xrightarrow{{}^{\cdot}OCF_3} F_2NOCF_3$$

$$FC(O)(CF_2)_3C(O)F + 2CsF \longrightarrow Cs^+{}^-OCF_2(CF_2)_3CF_2O^-Cs^+ \xrightarrow[-e]{F_2} Cs^+ {}^-OCF_2(CF_2)_3CF_2OF$$

$$\xrightarrow{-CsF} {}^{\cdot}OCF_2(CF_2)_3CF_2O^{\cdot} \xrightarrow{-COF_2} {}^{\cdot}OCF_2(CF_2)_3^{\cdot} \longrightarrow O\begin{array}{c} \diagup CF_2-CF_2 \\ | \\ \diagdown CF_2-CF_2 \end{array}$$

Thus the use of CsF to activate the carbonyl-containing reagents leads to significantly milder reaction conditions, but in most cases it fails to give the expected results due to the specific character of the processes.

Milder reaction conditions may be brought about by the presence of anionoid donor centres in the reactant systems. This is also the case when the starting covalent compounds contain reasonably polarised multiple bonds. This refers to thioanalogues of carbonyl-containing compounds. Thus thiophosgene spontaneously reacts with CF_3OF at temperatures as low as 195 K, forming two

Dioxolanes are the primary products of the reaction of $CF_2(OF)_2$ and hexa-fluorobenzene [160].

Under conditions of photochemical initiation, the yield of cyclohexene (9) is much higher than that of diene (8), as the process involves partial isomerisation of hexafluorobenzene to hexafluorobicyclo[2,2,0]hexadiene-2,5 (10). The latter is transformed by $CF_2(OF)_2$ to a mixture of mono- and dioxolanes (11–13) containing an insignificant amount of oxirane (14) [162,162].

In the thermal reaction of $CF_2(OF)_2$ with diene (10), compounds 11–14 are not formed because of the instability of the $FOCF_2O\cdot$ radical in these conditions.

Owing to the high reactivity and the presence of two fluorooxy groups, polymerisation of hexafluorobenzene and octafluoronaphthalene by $CF_2(OF)_2$ proceeds rather effectively. The structures of the polymer products are very complex and their representation requires much space. For example, polymer formed from C_6F_6 and $CF_2(OF)_2$ involves cyclohexadiene and cyclohexene fragments 15 and 16 [163].

The molecular mass of these polymers reaches 2500. These are low-melting point substances.

The example of perfluoropyridine shows that polymerisation by hypofluori-tes extends to practically the whole series of perfluoroaromatic carbons. With $CF_2(OF)_2$ pentafluoropyridine reacts in milder conditions than hexafluoro-benzene, giving a viscous liquid with molecular mass of 1000 to 2000, the yield of which is about 70% at the mole ratio of hypofluorite to C_5F_5N 1:3 [164].

17

Further treatment of compound *17* by hypofluorite leads to its partial fluorination and dioxomethylation with formation of product *18* containing about 18% of azacyclohexene (*19*) and azacyclohexadiene (*20*) as impurities, in the 1:3 mole ratio.

18   19   20

When the mole ratio of pentafluoropyridine to hypofluorite is raised to 1:1, the yield of products *19* and *20* increases to 75%.

The presence of a reactive functional group in the ring of perfluoroaromatic compounds imparts some specific features to the process of their interaction with hypofluorites. Thus the products of the reaction of CF_3OF with pentafluorophenol are perfluorinated cyclohexadienones and phenoxycyclohexadienone [165].

As in the reaction of C_6F_5OH with hydrogen peroxide, the key intermediate here seems to be the pentafluorophenoxyl radical [166].

A rather interesting case unambiguously supporting the radical nature of the reaction of perfluorinated unsaturated compounds with hypofluorites was described in patent [167]. Trifluoromethyl hypofluorite was suggested for use as an initiator of oxidation of perfluoroisopropylethylene by oxygen to polyoxydifluoromethylenecarbonyl fluorides.

$$(CF_3)_2CFCF{=}CF_2 + CF_3OF + O_2 \rightarrow CF_3O(CF_2O)_nCF_2COF$$

The by-product is perfluoroisobutyryl fluoride, which suggests formation of the respective alkyl radicals at the stage of initiation; these react with oxygen,

$$PhCH=CHPh + CF_3OF \xrightarrow{MeOH} PhCHF-CH(OMe)Ph$$

The most essential feature of these reactions is their occurrence in rather mild conditions (195 K). In this case *trans*-stilbene is transformed to a mixture of *threo*- and *erythro*-isomers in the ratio of 2:1, and *cis*-stilbene gives the same (but reverse) ratio of stereoisomers [182]. When *trans*-stilbene is treated with CF_3OF in the presence of Et_2O, the stereospecificity of addition increases to the ratio of *threo* to *erythro* 4:1, in the case of *cis*-stilbene the *erythro* to *threo* ratio becomes 3:1 [182].

$$PhCH=CHPh + CF_3OF \xrightarrow{Et_2O} PhCHFCH(OCF_3)Ph$$

The total yield of products of addition to *cis*-stilbene approaches 60%; to *trans*-stilbene, slightly exceeds 40%. In addition, the products of fluorination of double bond are formed in yields of 40 and 56% respectively.

$$PhCH=CHPh + CF_3OF \rightarrow PhCHF-CHFPh$$

It is interesting to note that the stilbene *trans*-isomer is in general more liable to fluorination. In the absence of polar solvents (in the $CFCl_3$ solution), the ratio of the addition to fluorination products for *trans*-stilbene is about 60:40; for *cis*-stilbene, 87:12.

The analysis of reasons, which brought about the hypothesis about the electrophilic nature of hypofluorites, shows that it has also been substantiated by stereospecificity of addition, apart from the above-mentioned *ortho*- and *para*-orientation of substitution of hydrogen. As suggested in [182, 193], the first stage of the reaction involves formation of the carbonium ion, whose further transformations determine the experimental conditions and the character of substituents at the double bond. In the inert solvents, the carbonium ion is stabilised by capturing the counterion CF_3O^-. The electron-donating solvents, such as Et_2O, stabilise the carbonium centre, giving difluoride. A similar effect is produced by the electron-donating substituents bonded with the carbonium centre. When the reaction is carried out in the presence of MeOH, the carbonium centre adds methoxide.

However many facts contradict these ideas. Thus, apart from the above-mentioned examples of formation of benzyl fluoride and tri-fluoromethoxy-benzene, the study of the reaction of stilbene with pentafluoroethyl hypofluorite is contradictory [197]. The results of the reaction of CF_3OF with 1,1-diphenyl-ethylene producing at least five different products cannot be explained in terms of the electrophilic mechanism [198].

$$Ph_2C=CH_2 + CF_3OF \xrightarrow[195\ K]{CH_2Cl_2} Ph_2C(OCF_3)CH_2F + Ph_2CFCH_2F$$

$$+ Ph_2C=CHF + Ph_2C(OCF_3)CHF_2 + Ph_2\overset{|}{C}CH=CPh_2$$
$$\overset{|}{CH_2F}$$

Explanation by this mechanism of the synthesis of α-fluoroketones from the respective vinyl acetates is very objectionable [199–201].

It is noteworthy that the results of the reactions of some of these compounds with CF_3OF and elemental fluorine are very similar. Thus, the reaction of fluorine with 1,1-difluoroethylene gives three out of five products—analogues of the reaction with CF_3OF [202].

$$Ph_2C=CH_2 + F_2 \rightarrow Ph_2CFCH_2F + Ph_2C=CHF + Ph_2CFCHF_2$$

The reaction of fluorine and diphenylacetylene in MeOH leads to the formation of the respective mono- and dimethoxy-derivatives [202].

$$PhC\equiv CPh + 2F_2 \xrightarrow{MeOH} PhCF_2CF_2Ph + PhCF(OMe)CF_2Ph$$

$$+ PhC(OMe)_2CF_2Ph$$

In the absence of methanol, only tetrafluorodiphenylethane is formed, which is the analogue of the product of the reaction between diphenylacetylene and CF_3OF [170,203].

$$PhC\equiv CPh + 2CF_3OF \rightarrow PhCF_2CF(OCF_3)Ph$$

The impossibility in principle to generate the fluorine cation by the heterolysis of the F–F bond under the action of unsaturated C=C compounds makes us reject the electrophilic mechanism of the reaction involving fluorine. However, since they are self-initiated at rather low temperatures, an alternative mechanism to account for the observed results seems to be the one-electron transfer. In this case it would be better to substitute the widely used term "electrophilic" by the term "electron affinity". The electron affinity of the fluorine molecule is known to

4.2   Group V Pentafluorides and Their Derivatives in Fluoroorganic Synthesis

Phosphorus and arsenic pentafluorides are gases, antimony and vanadium pentafluorides are viscous associated liquids, whereas niobium, tantalum and bismuth pentafluorides are solids (Table 1). All of them are vigorously hydrolysed, and are converted by the reducing agents to stable trifluorides, excluding vanadium which forms the stable paramagnetic tetrafluoride apart from trifluoride.

4.2.1   Antimony Pentafluoride and Fluoroantimonates

Antimony pentafluoride is a strong fluorinating agent. It should be borne in mind that it is also one of the most powerful Lewis acids, whose presence may give rise to other processes apart from fluorination (see, e.g. [7,8]). It acts as a strong fluorinating agent in the reactions of chlorine (bromine, iodine) exchange for fluorine at a saturated carbon atom, and in the oxidative fluorinations of multiple bonds in polyhalogenated alkenes.

The processes of halogen exchange for fluorine at a saturated carbon atom in the reactions with SbF_5 or $SbCl_5$–HF, and the application of these reactions in organic synthesis are described in [1]. We shall give some examples to illustrate these transformations. It should be noted that the reactions of SbF_5 afford fluorine-containing alkanes, ketones, and ethers from polyhalogenated derivatives, but, as a rule, it is impossible to have all halogens replaced by fluorine.

Heating of 1-H-4-chloroperfluorobutane with SbF_5 at 175 °C for 3 h leads to the formation of 1-H-perfluorobutane with a 50% yield [9]. Boiling of methyl pentachloroethyl ether with antimony pentafluoride yields methyl (α,α-difluorotrichloroethyl) ether (84%), with residual chlorine atoms unsubstituted

Table 1. Properties of group V pentafluorides [6]

Fluoride	Mol. wt.	M.p. (°C)	B.p. (°C)	Density (g cm^{-3}) (°C)	Heat of formation (kJ mol^{-1})
PF_5	126	−93.7	−88.6	5.805(g l^{-1})	−1593
AsF_5	170	−79.8	−52.8	2.33(−52.8)	−1237
SbF_5	217	8.3	142.7	2.99(23)	−1379 [5a]
BiF_5	304	151.4 [5b]	230 [5b]	5.52(25)	
VF_5	146	19.5	48.0	2.18(19)	−1481
NbF_5	188	79.5	234.5	3.29(25)	−1814
TaF_5	276	96	229.2	4.98(15)	−1828

by fluorine [10]. At the same time, Dear [11] reported fluorination of β,β,β-trichlorohexafluoro-*tert*-butanol by antimony pentafluoride, giving perfluoro-*tert*-butanol with a 92% yield.

$$CHF_2CF_2CF_2CF_2Cl + SbF_5 \xrightarrow[175\,°C,3h]{} CHF_2CF_2CF_2CF_3$$
$$50\%$$

$$CCl_3CCl_2OCH_3 + SbF_5 \xrightarrow[140\text{ to }150\,°C]{} CCl_3CF_2OCH_3$$
$$84\%$$

$$CCl_3C(CF_3)_2OH + SbF_5 \xrightarrow[35\text{ to }65\,°C]{} (CF_3)_3COH$$
$$92\%$$

Under the action of SbF_5 iodine is replaced by fluorine in iodo-perfluoroalkanes; here the efficiency of fluorination by SbF_5 is close to that in the reactions with chlorine and bromine trifluorides [12].

$$CF_3(CF_2)_nCF_2I + SbF_5 \rightarrow CF_3(CF_2)_nCF_3 \quad (n = 1\text{--}8)$$
$$(60\text{--}70\%)$$

It is worthwhile mentioning the use of antimony pentafluoride as a fluorooxidant. First it was found in the reactions of SbF_5 with polychlorinated alkenes which are transformed to polyfluoroalkanes. Thus on heating with SbF_5, hexachlorobutadiene forms 2,3-dichlorohexafluoro-2-butene and a small amount of 2-chloroheptafluoro-2-butene [13].

$$CCl_2=CCl-CCl=CCl_2 + SbF_5 \rightarrow CF_3CCl=CClCF_3 + CF_3CF=CClCF_3$$

In a similar way proceeds fluorination of hexachlorocyclopentadiene by antimony pentafluoride, leading to 1,2-dichlorohexafluorocyclopentene in a 50% yield [14,15]. In contrast to this, octachloro-2,5-diaza-1,5-hexadiene reacts with SbF_5 at 20 °C with the substitutive fluorination and cyclization. Heating of the resulting product with antimony pentafluoride leads to the substitution of the residual chlorine atom and formation of the immonium salt [16].

$$CCl_2=N-CCl_2CCl_2-N=CCl_2 \quad + \quad SbF_5 \xrightarrow[\substack{20°C, \\ 3\text{ days}}]{SO_2} F_3C-N \overset{Cl}{\underset{F}{\diagdown}} N \xrightarrow[60°C]{SbF_5} F_3CN \overset{+}{\underset{F}{\diagdown}} N-SbF_5^-$$
$$82\%$$

The polyfluorinated ethylene derivatives C_2F_3X react with SbF_5 at the atmospheric pressure with addition of two fluorine atoms, forming the respective polyfluoroethanes [17].

$$CF_2=CFX + SbF_5 \xrightarrow[50\,°C]{} CF_3CF_2X \quad (X = H, F, Cl)$$

Perfluoropropylene is fluorinated by SbF_5 at 70 to 80 °C (pressure) giving perfluoropropane (44% yield) but at the same time the dimer $(CF_3)_2CFCF=CFCF_3$ is formed (26% yield) [18]. The reaction of 1,1-difluoroethylene with antimony pentafluoride (50 °C) unexpectedly gave tris (β,β,β-trifluoroethyl)difluorostibine [17]. In the presence of catalytic amounts of SbF_5, the higher terminal polyfluoroalkanes are isomerised to the respective internal alkenes without fluorination [19].

Perhalo derivatives of cyclohexene are much more stable towards SbF_5 than fluorine-containing ethylenes and do not react with it at temperatures below 100 °C. At 140 to 150 °C 4-bromo- and 1-iodononafluorocyclohexenes add two fluorine atoms to form bromo(iodo)undecafluorocyclohexanes [20]. It is interesting that 1-chorononafluorocyclohexene disproportionates in the same conditions to 1,2-dichlorooctafluorocyclohexene and perfluorocyclohexene.

On heating with antimony pentafluoride at $\geqslant 100$ °C, polyhaloaromatic compounds are fluorinated to the respective derivatives of cyclohexene. This was first found by McBee with co-authors [14], when they treated hexachlorobenzene with SbF_5 at 125 °C, which resulted in 1,2-dichlorooctafluorocyclohexene. Later it appeared that, apart from this compound, the reaction gives 1,2,4-trichloroheptafluorocyclohexene, 1,2,4,4-tetrachlorohexafluorocyclohexene, and traces of 1,2-dichlorohexafluorocyclopentene [21].

According to [22], antimony pentafluoride reacts in a similar way with perchloronaphthalene and perchlorodiphenyl, but the structure of the products of fluorination was not determined.

Upon heating with SbF_5 in a sealed tube, the pentafluorobenzene derivatives C_6F_5R (R = F, Cl, Br, I) are fluorinated to form mainly 1-R-nonafluorocyclohexenes [23]. Apart from them, the reaction products contain perfluorocyclohexene, R-undecafluorocyclohexane (R = Br, I) and 1,2-dichlorooctafluorocyclohexene (R = Cl). Pentafluorophenol reacts with SbF_5 at 100 °C in an open system to give perfluoro-2-cyclohexenone, whereas pentafluoroanisole and pentafluoroaniline are completely resinified.

Octafluorotoluene and pentafluorobenzenesulfonyl fluoride are not fluorinated by antimony pentafluoride even after prolonged heating at 140 to 160 °C in a sealed tube.

Perfluorinated diphenyl [24] and naphthalene [24,25] react with SbF$_5$ upon heating, with addition of only four fluorine atoms.

2-R-Heptafluoronaphthalenes (R = H, Cl, CF$_3$) treated with antimony penta-fluoride are converted to perfluorotetralin and 6-R-perfluorotetralins (R = H, Cl) [24]. But heating of 2-bromoheptafluoronaphthalene and SbF$_5$ unexpectedly led mainly to perfluoro-1-methylindan [26]. As this is also the product of the reaction 2-bromoundecafluorotetralin with SbF$_5$ [27], the transformation of polyfluoronaphthalenes to the respective tetralins under the action of SbF$_5$ may be regarded as the common property of these compounds

For the reaction of SbF$_5$ with perfluoromethylindans, there has been found an interesting dependence of fluorination route on the position of the CF$_3$ group. It appeared that fluorination of perfluoro-2-methylindan occurs only in the aromatic ring, whereas fluorination of other isomeric methylindans (as well as perfluoroindan) leads to cleavage of the aliphatic ring [28].

The same authors reported that in the presence of the catalytic amounts of bromine or iodine, perfluorinated indan, 4- and 5-methylindans react with SbF_5 to form perfluoro[4.3.0]bicyclo-1-nonene. From perfluorotetralin they obtained perfluoro[4.4.0]bicyclo-1-decene. The reactions were proved to proceed via the initial bromofluorination of the aromatic ring with subsequent fluorodebromination.

In the reaction of perfluorobenzocyclobutene with Br_2–SbF_5 the only product is 2-bromoperfluoroethylbenzene.

The aromatic hydrocarbons are not fluorinated by antimony pentafluoride, but are resinified for the most part. In the case of the reactions of SbF_5 with benzene derivatives, it has been shown that, apart from resinified products, the reaction mixture contains arylstibines [29].

Antimony pentafluoride may be used for fluorinations not only at the carbon atom. For example, pentafluorobenzenesulfonyl chloride treated with SbF_5 (20 to 25 °C) is quantitatively converted to pentafluorobenzenesulfonyl fluoride [30]. Pentafluorobenzenesulfenyl chloride and pentafluorophenyldifluorophosphine are oxidised by SbF_5 to form the salts of the pentafluorophenyldifluorosulfonium and pentafluorophenyltrifluorophosphonium cations respectively [31,32]. On heating with SbF_5 phenyldichlorophosphine yields phenyltetrafluorophosphorane [33]. An example of the use of antimony pentafluoride to obtain fluorosilanes from chloro derivatives is the transformation of β-chloro-α,α-difluoroethyltrichlorosilane to β-chloro-α,α-difluoroethyltrifluorosilane [34]. It is interesting that interaction of tris-*tert*-butylsilane with SbF_5 also leads to the substituted fluorosilane [35].

Milder fluorinating agents are phenyltetrafluorostibine and diphenyl trifluorostibine. They provide smooth substitution of chlorine by fluorine in benzotrichloride and phenylpentachloroethane without involving the aromatic ring [36].

4.2.2 Vanadium Pentafluoride

Despite the fact that vanadium pentafluoride has been known for more than 30 years, its chemical properties have been little studied. In contrast to other group V element pentafluorides, this fluoride was found to be a very weak Lewis acid, but a substantially stronger fluorooxidant. Vanadium pentafluoride is easily soluble in HF, SO_2FCl, $CFCl_3$, ClF_3, BrF_3, BrF_5, SbF_5, Cl_2, Br_2 but has poor solubility in cyclo-C_4F_8, perfluorohexane, SF_6, AsF_5 [37,38]. With tetrahalomethanes CX_4 vanadium pentafluoride reacts at 20 °C, forming CF_4, CF_3X (X = Cl, Br, I), and CF_2Cl_2 (X = Cl) [39]. Carbon disulfide is transformed to bis(trifluoromethyl) di- and trisulfide and sulfur tetrafluoride, with VF_5 reduced to the trifluoride [40]. Heptane and cyclohexane reduce vanadium pentafluoride to the tetrafluoride, but the structure of the organic products formed is unknown [41]. With pyridine VF_5 reacts vigorously even at $-78\,°C$, forming a dark-brown precipitate insoluble in pyridine excess [42]. Under the action of ammonia, ethylenediamine, and tetramethylethylenediamine, VF_5 is reduced without fluorination of organic compounds [42].

Recently there have started the systematic studies of fluorination of poly-halogenated unsaturated and aromatic compounds by vanadium pentafluoride. Fluorination of terminal polyfluoroalkenes readily proceeds at -20 to $-30\,°C$ in $CFCl_3$ to give fluorine-containing alkanes [43].

$$CF_2=CXR_F + VF_5 \rightarrow CF_3CFXR_F$$

$$(X = Cl, R_F = C_2F_5, X = F, R_F = (CF_2)_4CFClCF_2Cl)$$

In a similar process tetrachloroethylene is transformed to 1,2-difluorotetra-chloroethane (72%).

Internal perfluorinated alkenes react with VF_5 under more severe conditions. Thus the transformation of isomeric perfluoromethylpentenes to perfluoro-2-methylpentane requires heating them with a 3-fold excess of VF_5 at 100 to 150 °C, but even under these conditions conversion of perfluoroalkenes is low. At the same time, replacement of vinyl fluorines by chlorine apparently facilitates fluorination of the C=C bond by VF_5, as 2,3-dichlorohexafluoro-2-butene is transformed to the respective alkane in the reaction with VF_5 at 20 to 30 °C. The effect of the nature of substituent on the rate of fluorination is more pronounced in the case of perfluoromethyl vinyl ether: unlike other terminal alkenes, this ether is fluorinated slowly even at 40 to 50 °C.

Vanadium pentafluoride easily fluorinates polyhalogenated 1,3-dienes [43]. At -20 to $-30\,°C$ 2-chloroperfluoro-1,3-butadiene is converted to cis-, trans-2-chloroperfluoro-2-butene and 2-chloroperfluorobutane (yields 56 and 8% respectively). As the alkene formed does not further react with VF_5 in these conditions, the polyfluorinated alkane is presumably the product of fluorine 1,2-addition to the starting fluorine-containing diene and subsequent fluorination of terminal butene.

$$CF_2=CCl–CF=CF_2 + VF_5 \xrightarrow{\text{CFCl}_3} CF_3CCl=CFCF_3 + CF_3CFClCF_2CF_3$$

Perchlorinated 1,3-butadiene in the same conditions adds 4 fluorine atoms, forming nearly equal amounts of hexachloro-1,2,3,4-tetrafluorobutane and hexachloro-1,1,2,4-tetrafluorobutane. Prolongation of the reaction from 0.5 to 3 h does not alter the isomer ratio.

As for tetrachloroethylene, there is no exchange of chlorine atoms for fluorine. It is not clear as yet, whether hexachloro-1,1,2,4-tetrafluorobutane is the primary product or is obtained from another isomer.

Polyfluorinated cycloalkenes are fluorinated by VF_5 in more rigid conditions than the terminal polyfluoroolefins, and their reactivity approaches that of the

The interaction of VF_5 with alkylpentafluorobenzenes proceeds in a more complex manner. Not only the benzene ring is fluorinated, but also the α-hydrogen atom is substituted by fluorine even in heptafluorotoluene [47].

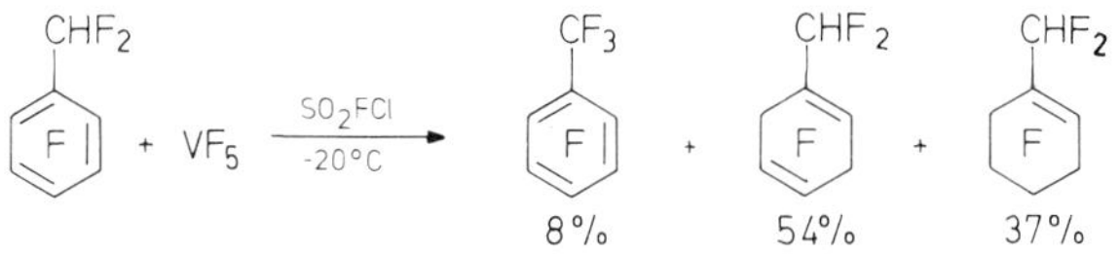

Thus vanadium pentafluoride is a promising fluorinating agent characterised by the fluorinating and oxidative properties. The data available suggest that the relative reactivity of polyfluorinated compounds decreases in the series:

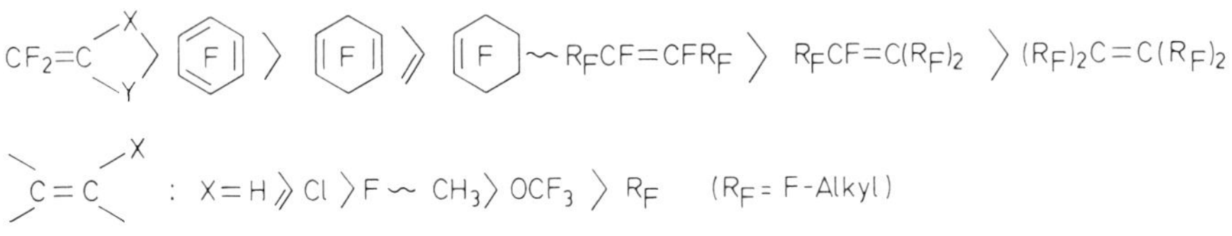

In this case the chlorine, bromine and iodine atoms, and the $OAlk$, NO_2, and CN groups do not hinder fluorination of the polyfluorinated unsaturated and aromatic compounds and are not changed by VF_5. There is no cleavage of the carbon–deuterium bond in the fluorination of deuteropentafluorobenzene by VF_5 [45], nor of the carbon–hydrogen bond in alkoxypentafluorobenzenes [47], but the alkyl hydrogen atoms in the allyl or benzyl position may be substituted by the fluorine atoms.

The fluorinations are carried out by adding VF_5 in a solvent or without it to the solution of a substrate in SO_2FCl, $CFCl_3$ or CH_3CN. Hydrogen fluoride readily dissolves VF_5, but is less suitable because of the low solubility of polyhalogenated organic compounds in it. Suitable reactors are of quartz, polytetrafluoroethylene, polychlorotrifluoroethylene (Kel-F), stainless steel, nickel, or copper. When necessary (at a temperature above $50\,^\circ$C), the process is conducted in a steel or nickel tube with a copper or Teflon seal, but in this case the reaction mixture should be shaken or stirred because of the formation of insoluble vanadium tetrafluoride. The fluorination products are easily separated either by pouring them onto ice after hydrolysis of the reaction mixture, or by distilling them off from the reactor. In the latter case the residual VF_4 may be converted to VF_5 by the reaction with fluorine at 180 to $220\,^\circ$C.

4.2.3 Phosphorus Pentafluoride and Fluorophosphoranes

Literature contains few data on the use of phosphorus pentafluoride for fluorination of organic compounds. Phosgene was reported [49] to be converted by PF_5 under rigid conditions (temperature, pressure) to dichlorodifluoro-methane. Tris (*tert*-butyl)silane treated with PF_5 in 1,1,2-trichlorotrifluoro-ethane is transformed to tris(*tert*-butyl)fluorosilane [50]. Interaction of tetraphenyltin with PF_5 in an autoclave leads to the formation of phenyltetra-fluorophosphorane and triphenyltin hexafluorophosphate. Heating of the latter yields fluorotriphenyltin [51].

$$(C_6H_5)_4Sn + PF_5 \xrightarrow[135\,°C,\ 20\ h]{} C_6H_5PF_4 + (C_6H_5)_3SnPF_6$$

$$(C_6H_5)_3SnPF_6 \xrightarrow{\Delta} (C_6H_5)_3SnF + PF_5$$

With phenol phosphorus pentafluoride forms phenoxyphosphoranes $(PhO)_nPF_{5-n}$ $(n = 1 - 3)$ and $(PhO)_4P^+PF_6^-$, but no fluorination of the aromatic ring occurs [52].

There are reports of the fluorination of the benzaldehyde carbonyl group with diphenyltrifluorophosphorane, leading to benzal fluoride (moderate yield) [52]. Phenyltetrafluorophosphorane vigorously reacts with aldehydes and ketones, forming fluorine-free polymer products. Hexamethyldisiloxane reacts with $PhPF_4$ to give fluorotrimethylsilane (yield (93%), and propionic anhydride forms propionyl fluoride (yield 91%) [53]. Succinic anhydride is transformed to succinyl difluoride in more rigid conditions (120 to 140 °C, yield 80%), but maleic anhydride polymerises in the reaction with $PhPF_4$.

Alcohols react with phenyltetrafluorophosphorane, forming organic fluori-des and olefins in low yields. However, if, instead of alcohols, $PhPF_4$ is made to react with their trimethylsilyl ethers, the reaction smoothly leads to phenyl-alkoxytrifluorophosphoranes, which decompose upon heating with liberation of alkyl fluorides [54–56].

$$ROH + Me_3SiCl \rightarrow ROSiMe_3 \xrightarrow{PhPF_4} ROPF_3Ph \rightarrow RF + PhPOF_2$$

$$R = Me,\ Et,\ i\text{-}Pr,\ MePrCH-,\ cyclo\text{-}C_6H_{11},\ t\text{-}Bu$$

Phenol and 2,2,2-trichloroethanol form very stable phosphoranes $ROPF_3Ph$ $(R = Ph, CCl_3CH_2)$; their attempted transformation to the respective fluorides by heating to 150 to 200 °C failed.

For the direct transformation of alcohols to fluorides, the use of triphenyldi-fluorophosphorane [57] and diphenyltrifluorophosphorane [58] has been sug-gested. The syntheses are carried out at 130 to 180 °C in acetonitrile for 6 to 10 h. Yield of the alkyl fluoride depends on the structure of the alkyl group: primary aliphatic alcohols are transformed to alkyl fluorides in 50–80% yields; secondary alkyl fluorides are formed with lower yields, and cyclohexanol gives only cyclohexene.

$$ROH + Ph_2PF_3 \xrightarrow{MeCN} RF + Ph_2POF + HF$$

$$R = C_5H_{11}(62\%),\ C_8H_{17}(76\%),\ PhCH_2(32\%),\ PhCH_2CH_2(52\%),$$
$$MePrCH(54\%),\ CH_2ClCH_2CH_2(64\%)$$

Alcohols and acids do not react with MoF_6 in these conditions.

It should be noted that in the absence of a catalyst, molybdenum hexafluoride fluorinates ketones only at temperatures above $100\,°C$ [71].

Molybdenum hexafluoride has been used to obtain aryl trifluoromethyl ethers from aryl chlorothioformiates [72]. For that purpose the reagents are mixed at $-25\,°C$ in a metal reactor and heated. Yields of the target products are 70–90%.

$$RC_6H_4OH + Na \rightarrow RC_6H_4ONa \xrightarrow{CSCl_2} RC_6H_4OCSCl$$

$$\xrightarrow{MoF_6} RC_6H_4OCF_3$$

$$R = H(40\%),\ 2\text{-Me}(70\%),\ 3\text{-Me}(61\%),\ 4\text{-Me}(87\%),\ 4\text{-Cl}(90\%),$$
$$3\text{-F}(66\%),\ 4\text{-F}(95\%),\ 3\text{-CF}_3(70\%),\ 4\text{-Br}(89\%)$$

Chloroformiates are much more difficult to fluorinate with molybdenum hexafluoride. For example, phenyl chloroformiate was only transformed to phenyl fluoroformiate in a 30% yield, and there is no substitution of the carbonyl oxygen by fluorine. It means that the C=S group is more reactive toward MoF_6 than the C=O group.

Under the action of MoF_6 in mild conditions, the acyl and aroyl chlorides and the respective carboxylic acids are only transformed to the acyl (or aroyl) fluorides [71–77]. Dissolution of MoF_6 in acetic acid proceeds with evolution of heat, the temperature of the solution rising to 40 to $60\,°C$, to form acetyl fluoride (quantitative yield) and a colourless solution of $MoOF_4$ in acid excess [75]. If the fluorination is conducted at 130 to $140\,°C$ in an autoclave, the products are the respective 1,1,1-trifluoroalkanes with 60–80% yields. The chlorine and bromine atoms in α-position to the carboxyl group remain intact in this process, but the dichloro- and difluoroacetic acids undergo decarboxylation (or decarbonylation) leading to the methane derivatives [73].

$$CH_2RCOOH + MoF_6 \rightarrow CH_2RCF_3$$

$$R = H(63\%),\ Cl(88\%),\ Br(89\%)$$

$$CF_3COOH + MoF_6 \xrightarrow{130\,°C,19\,h} CF_3COF$$

$$CHCl_2COOH + MoF_6 \rightarrow CHCl_2CF_3 + CHClF_2 + CHF_3 + CO + CO_2$$

The latter result is in disagreement with the data of [75], where the difluoroacetic acid and MoF_6 are reported to react for 20 h at $190\,°C$ to form pentafluoroethane with a 60% yield.

Aliphatic esters react with molybdenum hexafluoride under the same conditions as acids. For example, ethyl acetate is transformed to 1,1,1-trifluoroethane with a 46% yield upon heating with two equivalent of MoF_6 at 130 °C for 16.5 h [73].

Fluorination of the aromatic carboxylic acids is illustrated by the reaction of the 3- and 4-pyridinecarboxylic, and 2,6-pyridinedicarboxylic acids [75]. Substitution of oxygen by fluorine proceeds in rigid conditions, requiring the use of 3.5 to 5 mole of MoF_6 per mole of the acid.

The reaction is performed in an autoclave, the yield of the trifluoromethylpyridines being 60–80%.

The transformations of aroyl chlorides to the substituted benzotrifluorides under the action of molybdenum hexafluoride proceed more readily than of the acids themselves [73]. Aroyl chloride and MoF_6 are mixed at −20 °C in the ratio of 3:1 (mol), and the mixture is heated to 130 °C. The yields of the target products strongly depend on the substituent position in the ring, the *ortho*-substituted benzotrifluorides always obtained in low yields.

R - 2-F (1%), 4-F (55%), 2-Cl (1%), 4-Cl (17%), H (42%)

The presence of electron-accepting substituents in the aromatic ring facilitates fluorination of the chlorocarbonyl group, whereas the electron-donating substituents considerably slow down the process. For example, 4-methylbenzoyl chloride was converted to 4-methylbenzotrifluoride with an 8% yield.

Thiobenzoyl chloride reacts with MoF_6 in the same manner as benzoyl chloride; the yield of benzotrifluoride in both cases is close to 40% [73].

Puy [71] has shown that molybdenum hexafluoride at 130 °C transforms 2,2,2-trifluoroethanol to 1,1,1,2-tetrafluoroethane (85%), and ethylene oxide gives 1,1-difluoroethane with a 42% yield. The reaction of MoF_6 with dimethyl sulfite has been reported, where the products are methyl fluoride and methyl fluorosulfinate [78]. There are no other examples of fluorination of these functional groups in literature, therefore it is early to judge how general these transformations are. It should be noted that, unlike carbonyl compounds, carbonyl hydrazones react with molybdenum hexafluoride without formation of fluoroorganic products [79].

Alcohols are dehydrated by UF_6 to the aldehydes, which are subsequently fluorinated to acyl fluorides [92].

$$ArCRHOH + UF_6 \rightarrow ArCOR + ArCOF$$

Ar	R			
C_6H_5	H	1	:	2(40%)
$4\text{-}CH_3C_6H_4$	H	1	:	2(38%)
C_6H_5	C_6H_5			(45%)
C_6H_5	CH_3			(64%)

Oximes, hydrazones, tertiary amines R_2CHNMe_2 and ethers R_2CHOMe are converted by uranium hexafluoride to the carbonyl compounds. In a similar way occurs cleavage of carboxylic acid hydrazones [91,92].

$$R_2C=N-NXY + UF_6 \xrightarrow[0\,°C,\ 2\,h]{C_2F_3Cl_3} (R_2C=N-\overset{+}{N}XY)F^- \xrightarrow{H_2O} R_2C=O$$
$$\underset{UF_5}{\big|} \qquad 50\text{--}96\%$$

$$R_2CHOCH_3 + UF_6 \xrightarrow[25\,°C,\ 1\,h]{C_2F_3Cl_3} R_2C=O$$

$$RCONHNH_2 + UF_6 \xrightarrow[45\,°C]{} RCOOH \quad (R\text{---alkyl, aryl})$$

Trimethylchlorosilane has been shown [93] to react with uranium hexafluoride at a temperature as low as $-78\,°C$ in $CFCl_3$ to give trimethylfluorosilane. This observation is interesting in that it demonstrates the dependence of the Si–Cl bond reactivity on other substituents at the silicon atom, since UF_6 reacts with $SiCl_4$ at approx. $25\,°C$ [94]:

4.4   Practical Recommendations and Preparations

In conclusion we would like to give some practical recommendations for the use of the penta- and hexafluorides considered in this chapter in the synthesis of fluoroorganic compounds to facilitate the choice of a suitable fluorinating agent.

Antimony pentafluoride is a convenient reagent for the synthesis of polyfluoroalkanes from haloperfluoroalkanes R_FX ($X = Cl$, Br, I) by the exchange of X for F at 60 to $170\,°C$. The aliphatic radical R_F may contain one or two hydrogen atoms, and the functional groups C–O–C, and $>C=O$, but the hydrocarbon analogues are resinified in the reaction with SbF_5.

Antimony and vanadium pentafluorides are strong fluorooxidants, VF_5 being more reactive than SbF_5. Using these fluorides, it is possible to transform polyhaloaromatic compounds to the polyfluorinated cyclohexadienes, cyclohexenes, and cyclohexanes, whereas polyhaloalkenes and -cycloalkenes may be converted to the respective alkanes (phosphorus fluorides, as well as molybdenum, tungsten, and uranium hexafluorides are not suitable for that). It is important to know that in the processes of oxidative fluorination by antimony pentafluoride, the exchange of the Cl (Br, I) atoms for fluorine is possible, but this does not occur in the reactions involving VF_5. Fluorination by VF_5 proceeds in far milder conditions than fluorination by SbF_5.

The phenyl-substituted phosphorus (V) derivatives (except for Ph_4PHF_2) and hexafluorides MoF_6 and WF_6 may be used for substituting oxygen by fluorine in the functional groups COOR, C=O, and C–OH. The most suitable reagent for this is molybdenum hexafluoride, which is a rival to SF_4 and fluorosulfuranes. Currently MoF_6 stands second (after SF_4) in importance as an agent for the transformation of the carbonyl and carboxyl groups to the difluoromethylene and trifluoromethyl groups respectively. The aromatic aldehydes and ketones react with MoF_6 in the presence of BF_3 to form geminal difluorides with an approx. 50% yield. The presence of substituents F, Cl, Br, NO_2 in the ring does not prevent fluorination. Molybdenum hexafluoride is more selective than SF_4, therefore arylketones may be fluorinated by it without involvement of groups COOR, $CONR_2$, CN, and POR_2. The ortho-substituted arylketones do not react with MoF_6. Molybdenum hexafluoride is a convenient reagent for the synthesis of aryl trifluoromethyl ethers from aryl chlorothioformiates.

Primary and secondary alcohols may be easily converted to alkyl fluorides using Ph_2PF_3 or Ph_3PF_2, but for the fluorination of carbonyl compounds they are unsuitable.

Fluorination of oxygen-containing compounds by WF_6 proceeds in relatively severe conditions and gives small yields. Hence this fluoride may hardly be recommended for the synthesis. At the same time, UF_6 allows to obtain acyl fluorides from alcohols and aldehydes in one step.

1. Fluorination with SbF_5

a) *Perfluoroheptane* [12]. 1-Iodoperfluoroheptane (2 g) was placed with an excess of SbF_5 in a small autoclave, which was heated at 250 °C for 8 h, then gradually to 320 °C for 24 h. Distillation gave perfluoroheptane (66% yield).

A similar procedure yields perfluoropropane (67%), perfluorobutane (70%), perfluoropentane (65%), perfluorohexane (68%), and perfluorodecane (66%).

b) *Perfluorotetralin* [25]. Octafluoronaphthalene (1.60 g, 5.9 mmol) and SbF_5 (5.60 g, 25.8 mmol) were heated for 15 min at 100 °C with intermittent shaking of the flask. Then the mixture was heated to approx. 250 °C (the temperature of a bath) for 0.5 h, with simultaneous distilling off of the product into a trap containing cold water. The organic layer was washed with water and dried over $MgSO_4$ to give 1.45 g (71%) of perfluorotetralin (b.p. 163 to 165 °C).

2. Fluorination with VF_5

a) *1,2-Dichlorodecafluorocyclohexane* [20]. In a 10 ml nickel or steel tube was placed VF_5 (7.0 g, 48 mmol), the tube was cooled to -10 to -20 °C, and 1,2-dichlorooctafluorocyclohexene (3.0 g, 10.1 mmol) was added. The tube was stoppered and shaken at 50 to 60 °C for 3 h. The reaction mixture was cooled to -10 °C, then poured onto ice, the organic layer was washed with water, dried over $MgSO_4$ and distilled. The yield of 1,2-dichlorodecafluorocyclohexane was 95%.

amounts of chlorine and undiluted fluorine. ClF prepared by this method may be further used in organic reactions without preliminary purification. Work with chlorine monofluoride is much more convenient if the ClF current is produced by passing the chlorine and fluorine under excessive pressure but not by evacuation of the reactor at the outlet, as suggested in [14].

On a laboratory scale, chlorine monofluoride may also be obtained by passing the equimolar quantities of chlorine and chlorine trifluoride through a nickel tube at 300 to 400 °C [4,6].

$$Cl_2 + ClF_3 \xrightarrow[\text{300 to 400 °C}]{} 3ClF$$

The unchanged ClF_3 and Cl_2 are separated from ClF by condensation in a copper trap cooled to -78 °C. But most frequently synthesis according to the above scheme is carried out by heating a mixture of Cl_2 and ClF_3 for several hours in a stainless steel autoclave at 150 to 200 °C (40 atm) [15]. The autoclave is pre-passivated by chlorine trifluoride.

Preparation of chlorine monofluoride from ClF_3 is convenient, as ClF_3 is commercially available and may be stored and transported in steel cylinders [1].

There are several laboratory methods for the preparation of ClF without using F_2 or ClF_3. One of them is based on the endothermic reaction of chlorine with metal fluorides, resulting in a mixture of gases containing Cl_2, ClF and ClF_3. Chlorine is slowly passed through a melt of the mixture of alkali metal fluorides, and ClF is isolated by freezing [16].

$$Cl_2 + MF \xrightarrow[\text{800 to 1000 °C}]{} ClF + MCl$$

With silver fluoride or $AgBF_4$ the reaction proceeds under milder conditions. Another route to ClF is the reaction of chlorine, SbF_5 and anhydrous HF at -78 to -35 °C [17].

Chlorine monofluoride is easily prepared by the reaction of chlorine fluoro-sulfate with well-dried CsF [2,18].

$$ClOSO_2F + CsF \rightarrow ClF + CsOSO_2F$$

Table 1 presents some data on the properties of chlorine monofluoride and other halogen fluorides.

Chlorine trifluoride is a low-boiling liquid of light green colour. Its industrial production methods, apparatus design and material have been described in detail [1,6]. ClF_3 is synthesized by passing the necessary quantities of chlorine and fluorine through a copper or nickel tube heated to 250 to 280 °C. Chlorine trifluoride is formed as a result of a sequence of reactions [11].

$$Cl_2 + F_2 \xrightarrow{k_1} 2ClF$$

$$ClF + F_2 \xrightarrow{k_2} ClF_3 \quad k_1 \gg k_2$$

Table 1. Some properties of halogens and halogen fluorides [1,4,23–25]

Compound	B.p. (°C)	M.p. (°C)	Dipole moment (D)	Atomic distances (Å)	E_{diss} (kJ mol^{-1})	Heat of formation (kJ mol^{-1})
F_2	−188	−219	—	1.418	156	—
Cl_2	−35	−101	—	1.988	240	—
ClF	−100	−155	0.89	1.63	253	−50.5
ClF$_3$[a]	12	−76	0.65	1.598–1.698	105[b]; 160	−164
ClF$_5$[c]	−13	−93	—	1.58; 1.67	—	−244
Br_2	59	−7	—	2.284	295	—
BrF	(20)[d]	−33	1.29	1.756	254; 260	−75.5
BrF$_3$[a]	126	9	1.19	1.72; 1.81	—	−256
BrF$_5$[c]	41	−30	1.51	1.69; 1.78	—	−429
I_2	184	113	—	2.666	147	—
IF	1 (decomp.)		—	1.909	277	—
IF$_3$[a]	−35 (decomp.)					
IF$_5$[c]	100	9	2.18	1.869; 1.844		−824; −840
IF$_7$	5 (subl.)	6			122[b]	

[a] Planar T-form structure
[b] The value of the heat of the reaction: $HlgF_n \rightarrow HlgF_{n-2} + F_2$
[c] Tetragonal pyramid
[d] The value was obtained by extrapolation

compounds is explained by the ability of these compounds to associate with formation of fluorine bridges [1,26].

A specific feature of halogen fluorides is their amphoterism [4]. Thus bromine trifluoride is capable of self-ionisation.

$$2BrF_3 \rightleftarrows BrF_2^+ BrF_4^- \rightleftarrows BrF_2^+ + BrF_4^-$$

For the same reason, halogen fluorides react both with Lewis acids and bases, forming salt-like compounds with complex fluorine-containing anions and cations [2,27,28].

$$XF_n + YF_m \rightarrow XF_{n-1}^+ YF_{m+1}^-$$

$$XF_n + MF \rightarrow M^+ XF_{n+1}^-$$

Chlorine monofluoride reacts with Lewis acids giving complexes of type $Cl_2F^+ YF_{m+1}^-$ [28], but not $Cl^+ YF_{m+1}^-$, as considered earlier [4,17], since Hlg^+ cations cannot exist in solution or salt-like compounds because of their thermodynamic instability [29]. Anhydrous HF is used, which also forms highly polar and slightly stable $1:1$ complexes with halogen fluorides. Halogen fluoride solutions in HF have a high electric conductivity at the expense of ionisation according to the scheme [1].

$$ClF_3 + HF \rightleftarrows ClF_3 \cdot HF \rightleftarrows ClF_2^+ + HF_2^-$$

In compounds with the complex fluorohalogenonium cation, the halogen atom bears a higher positive charge than in the starting molecule. Due to this, such compounds are even stronger oxidants than the starting halogen fluorides. The high electron affinity is demonstrated, e.g. by the oxidative fluorination of xenon by some complexes of this type [28,30].

$$Xe + 2ClF + AsF_5 \rightarrow XeF^+ AsF_6^- + Cl_2$$

$$3Xe + 2BrF_2^+ SbF_6^- + SbF_5 \rightarrow 3XeF^+ SbF_6^- + Br_2$$

Being strong oxidants, halogen fluorides can undergo reactions with most elements, as well as organic and inorganic compounds [1]. Of few materials that are stable against these extremely corrosion-active substances at moderate temperatures, it is necessary to mention nickel, copper, Monel (70% Ni, 30% Cu), Kel-F and Teflon. The reactor intended for work with halogens may also be made of quartz. Glass is slowly destroyed under the action of halogen fluorides, especially in the presence of moisture which easily decomposes all compounds of this class, forming HF. Liquid halogen fluorides, especially chlorine and bromine polyfluorides, are rather dangerous to handle [5,6]. Chlorine trifluoride ignites wood, plastics, and in contact with ordinary solvents at $-100\,°C$ causes an explosion. Bromine trifluoride is less dangerous, though at room temperature it reacts explosively even with chlorocarbons. Iodine pentafluoride is not so aggressive. Only polyfluorocarbons may be quite safely mixed with halogen polyfluorides at low temperatures. While performing the reactions with low-reactive hydrogen-containing substrates, accumulation of halogen fluorides in a mixture should be avoided. Inadherence to this rule may cause an explosion [5]. Safe work with chlorine and bromine polyfluorides is best provided by diluting an organic substrate with a solvent inert relative to halogen fluoride before bringing it in contact with these reagents. It should also be remembered that in their toxicity, halogen fluorides are similar to elemental fluorine.

5.3 Reactions of Bromine and Iodine Monofluorides

Bromine and iodine monofluorides are unstable compounds, therefore they have been little studied, but the reactions of their stoichiometric equivalents of "BrF" and "IF" have been studied quite intensively.

5.3.1 Stoichiometric Equivalents of "BrF" and "IF"

The precursors of "BrF" and "IF" are the systems: N-bromo- or N-iodo-substituted amide—anhydrous HF [5,8,10], more rarely—Hlg_2–HF or Hlg_2–AgF [31–35]. As shown below, this does not mean that bromine and iodine monofluorides are really formed in these systems. N-Halogen-substituted amides or molecular halogens are the source of electrophilic halogen, whereas HF or AgF give a nucleophilic fluoride ion.

These systems are usually used for the addition of "BrF" and "IF" at a multiple bond. As a rule, halofluorination proceeds at -80 to $25\,°C$ in solution [5,36]. The electron-donating substituents at a multiple bond accelerate the reaction, the electron-accepting ones slow it down [37]. Halofluorination by such systems presents typical electrophilic addition reactions defined by the general equation.

There are many preparative modifications of halofluorination, depending on the type of unsaturated compound and the source of electrophilic halogen. As a solvent, ether, tetrahydrofuran [5,36], HF(70%)–pyridine (30%) [31,38], or anhydrous HF [37] are used. Using oxygen-containing solvents is not always desirable, as they may participate in the addition reaction as nucleophiles [39].

Halofluorination in the HF-Py system gives good yields in the case of phenyl-substituted olefins [31,40,41]. Bromo- and iodofluorination of cyclohexene and its derivatives proceeds *anti*-stereospecifically [42,43]. The influence of substituents at a multiple bond on the regioselectivity and stereochemistry of bromofluorination of substituted alkenes has been well studied [8,10]. In general, the electrophilic and nucleophilic addition orientation is such as would be expected for the electrophilic mechanism of the reaction. As it is possible to subject unsaturated compounds with electron-deficient multiple bonds, which usually react with halogens via the free-radical mechanism in bromofluorination, it would be interesting to compare the competing effects of substituents on the addition orientation and stereochemistry of bromofluorination.

Bromofluorination of 1,3-dichloroprop-1-enes proceeds regiospecifically, and electrophilic bromine attaches to the central carbon atoms, whereas nucleophilic F^-—to the terminal carbon atom [44].

$$CHCl=CHCH_2Cl + {>}N{-}Br \xrightarrow{\;HF\;} CHClF{-}CHBrCH_2Cl$$

The chlorine atom at the multiple bond as a *p*-donor entirely controls the addition orientation. Introduction of substituents delocalising the positive charge at carbocation's 2-position reverses the addition orientation—electrophile is oriented to the terminal carbon atom, and nucleophile—to the central one [32].

$$CHCl=CRCH_2Cl \;+\; {>}N{-}Br \xrightarrow{\;HF\;} CHClBr{-}CRFCH_2Cl \qquad R=Me, Cl, F$$

The total orienting effect of the CH_2Cl with Me, F, and Cl substituents is stronger than that of chlorine.

Bromofluorination of acrylates also proceeds in accordance with the electronic interpretation of the Markownikoff rule. Therefore in the case of methyl acrylate, electrophilic bromine is predominantly oriented to the 2-, and for methyl methacrylate—to the 3-position [39,45]. Halogens as the *p*-donor substituents produce a much more stabilising effect on the intermediate carbocation than the CH_3-group [38,45]. Thus, bromofluorination of methyl *E*-2-methyl-3-chloropropenoate is regiospecific: the addition orientation is completely determined by the electronic influence of the chlorine atom.

The phenyl group is the π-donor and is a much more effective orientant than the halogen atom [40].

$$CHBr=CHPh \;+\; {>}N{-}Br \xrightarrow{\;HF\;} CHBr_2{-}CHFPh$$

Bromofluorination of multiple bonds proceeds chiefly stereospecifically, as *anti*-addition. In some cases this was strictly proved by the stereospecific dehydro-bromination of diastereomers to the respective fluorine-containing olefins [40,46]. The deviations observed in this case—incomplete *anti*-stereospecificity [32,40,46,47] or *syn*-addition [40]—may be explained in terms of the stability of bridged bromonium and open carbenium ions which are believed to be the intermediates in bromofluorination reactions. The incomplete *anti*-stereo-specificity of the addition of "BrF" to *E*-1,2-dichloroethylene and *E*-1,2,3-trichloropropene [32, 46] is attributable to the fact that the bromonium ions *1* with transoid chlorine atoms are less stable than ions *2* with cisoid chlorine atoms. This is in accordance with the greater stability of *cis*-dihaloethylenes than of *trans*-dihalaloethylenes [48,49].

R=H, CH$_2$Cl

This results in the partial rearrangement of ion *1* to ion *2*. That is why bromofluorination of *Z*-olefins is completely *anti*-stereospecific, whereas bromo-fluorination of *E*-olefins is only stereoselective (85–60% of *anti*-adducts respectively) [32,46].

The non-stereospecific addition of "BrF" to ethyl *E*- and *Z*-3-chloro-crotonates is explained by other reasons [47].

As a result of the addition of electrophilic bromine at a double bond of both *E*- and *Z*-3-chlorocrotonates, an intermediate is formed—the carbocation *3* rather than the bromonium ion. This occurs due to the presence of two stabilising substituents CH$_3$ and Cl at the carbocationic centre and the electron-accepting alkoxycarbonyl group. As a result, the reaction is regiospecific and non-stereospecific.

However it is enough to have at least a weak *p*-bridge bond in the non-symmetric bromonium ions *4, 5, 6* for the completely regiospecific reaction to proceed as an *anti*-stereospecific one [38,40,44,47].

The deviations from *anti*-stereospecificity of bromofluorination of phenyl-substituted ethylenes results from the fact that in the process of the reaction, *cis*-olefins isomerise to more stable *trans*-olefins [40]. In the case of the sterically hindered *cis*-2-phenyl-1-*tert*-butylethylene, there occurs 100% *cis*-addition, and for the *trans*-isomer—100% *anti*-addition.

The unsaturated acetylated glucals have been reported [50,51] to undergo iodo- and bromofluorination by N-iodo and N-bromosuccinimide in HF with formation of *syn*-adducts. As shown in [33], the ^{1}H and ^{19}F NMR spectral data indicate that these reactions proceed predominantly as *anti*-addition, so the configuration assignment of halofluorinated sugars in [50,51] is erroneous.

For the effective halofluorination of unsaturated sugars and other compounds that are easily polymerisable or liable to undergo rearrangements in HF, it is recommended to use the system: bromine (or iodine)—finely divided silver fluoride in benzene or the benzene—acetonitrile mixture [33,34].

Bromofluorination of norbornene and the related compounds by NBS and HF in ether or pyridine is accompanied by the Wagner–Meerwein rearrangement [31,52,53]. Norbornene gives three main products: 2-*exo*-fluoro-7-*anti*-bromonorbornane *7*, 2-*exo*-fluoro-5-*exo*-bromonorbornane *8*, and 3-fluoronortricyclane *9*.

The regioselectivity and stereochemistry studies have also been carried out for bromofluorination of substituted acetylenes by N-bromoacetamide in anhydrous HF [54]. This reaction affords vicinal bromofluoroalkenes in satisfactory yields, though treatment of substituted acetylenes with the Br_2–AgF system leads only to bromination products. Bromofluorination of terminal acetylenes proceeds regiospecifically and *anti*-stereoselectively (>95%).

5.3.2 Bromine and Iodine Monofluorides

The mixtures of stoichiometric quantities of bromine trifluoride and molecular bromine, or iodine pentafluoride and molecular iodine were used as a source of bromine and iodine monofluorides in the addition reactions of perfluoro- and fluorochloroolefins, and alkyl perfluoroalkenyl ethers [55–58].

On the basis of the analysis of the products of addition at the multiple bond of olefins and the fact that BrF was really found in the Br_2–BrF_3 mixture, the following simple schemes were suggested for these reactions [59].

$$Br_2 + BrF_3 \rightleftarrows 3BrF$$

$$2I_2 + IF_5 \rightleftarrows 5IF$$

$$\text{\}C=C\text{\{} + HlgF \rightarrow \text{\}CHlg - \overset{|}{C}F$$

However recent studies have shown that the reactions proceed in a more complex way, and in some cases halogen polyfluorides may themselves be the main reagents in such mixtures. Therefore these reactions will be considered in detail in Sect. 5.5.

The addition of halogen monofluorides IF and BrF at multiple bonds has been reported in [60–63]. Iodine and bromine monofluorides were obtained by passing molecular fluorine diluted with nitrogen (8–10% v/v) into the strongly diluted ($<1\%$) and cooled solutions of I_2 and Br_2 in the inert solvents ($CFCl_3$, $CHCl_3$). Addition of olefin to the resulting cool solution of HlgF gave the respective adducts with good yields. Some peculiarities of these reactions have been considered. Thus the reaction of olefins with iodine monofluoride in $CFCl_3$ easily proceeds according to the scheme.

$$C_6H_{13}CH=CH_2 \xrightarrow{\text{IF}} C_6H_{13}CHF-CH_2I$$
70%

Similar reactions with BrF proceed smoothly only in the presence of ethanol (traces) or any other alcohol. In the absence of alcohol, the reaction of BrF with olefins was very vigorous, and the bromofluoroadducts could not be isolated from the reaction mixture. The bromofluorination of olefins in the presence of an alcohol is suggested by the authors of [61] to proceed as follows.

$$BrF + ROH \xrightarrow{-75°C} ROBr + HF$$

The bromoalkoxy adducts were really obtained as admixtures among the fluorinated products. Bromofluorination of *cis*- and *trans*-stilbenes by this method proceeds *anti*-stereospecifically, whereas iodofluorination [60] involves further substitution of iodine in the adducts by fluorine. In this case, *cis*-stilbene

[44]. As in bromofluorination of these alkenes, the chlorine atom at a double bond controls the addition orientation: electrophilic chlorine is attached to the central carbon atom, and fluoride ion—to the terminal carbon. The addition of "ClF" to methyl E-2-methyl-3-chloro- and Z-2-fluoro-3-chloropropenoates also proceeds regioselectively and *anti*-stereospecifically [37,47]. The electrophilic chlorofluorination of ethyl E- and Z-3-chlorocrotonates proceeds non-stereospecifically due to the reason considered previously (see Sect. 5.3.1).

Using the systems involving chlorine donors and anhydrous HF, it is possible to carry out at −10 to 10 °C the substitutive fluorination of some organic bromoderivatives [20]. As a source of electrophilic chlorine, N-chloroamines (hexachloromelamine, trichloroisocyanuric acid) may be used, as well as some hypochlorites (trifluoroacetyl hypochlorite). The reaction follows the route.

$$-\overset{|}{\underset{|}{C}}Br \;+\; \overset{\diagdown}{\underset{\diagup}{N}}-Cl\,(or-OCl) \;+\; HF \;\longrightarrow\; -\overset{|}{\underset{|}{C}}F \;+\; \overset{\diagdown}{\underset{\diagup}{N}}-H\,(or-OH) \;+\; BrCl$$

Hypochlorites as a rule easily decompose in HF, leading to a decreased yield of fluoro-derivatives. NBS is unreactive in these conditions. For effective fluorination, the solubility of either the reagent or substrate in anhydrous HF is necessary. The best results (the yield of fluoro-derivatives 50–70%) have been obtained for polyhaloalkanes containing bromine at a secondary or tertiary carbon atom. The substitutive fluorination of 1,2-dibromo-3-chloroisobutane with hexachloromelamine–HF (one equivalent of N-chloroamine per one mole of haloalkane) gives exclusively 1-bromo-2-fluoro-3-chloroisobutane (*10*).

$$CH_2BrCBrCH_2Cl + R-NCl_2 \xrightarrow{\;HF\;} [BrCH_2\overset{+}{C}CH_2Cl] \xrightarrow{\;F^-\;} CH_2BrCFCH_2Cl$$
$$\quad\;\; \underset{Me}{|} \hspace{6.2cm} \underset{Me}{|} \hspace{3.4cm} \underset{\underset{10}{Me}}{|}$$

With an excess of electrophilic reagent, the primary bromine atom in alkane *10* is not substituted by fluorine. The rate-determining stage of the reaction is obviously the electrophilic elimination of bromine from the substrate, forming the intermediate carbocation which is stabilised by capturing the fluoride ion. The substitutive fluorination by these systems is not always regiospecific. The vicinal dibromides may give both fluoro-containing isomers, possibly due to formation of bridged bromonium ions.

When 1,2-dibromo-3-chloropropane *11* was treated with one equivalent of hexachloromelamine in HF, a 3:1 mixture of isomeric monofluorides *12, 13* was obtained with a total yield of 60%. With an excess of electrophilic reagent, only secondary bromide *13* is converted to difluorochloropropane *14*. As seen from the above examples of substitutive fluorinations, the chlorine atoms in the molecule remain intact.

This indicates that in the $>$N–Cl—HF systems, chlorine monofluoride is not formed. As will be shown (see Sect. 5.4.4), ClF easily substitutes bromine by fluorine in the primary bromides such as *10* and *12*, and in anhydrous HF it substitutes chlorine in the primary chloro-derivatives.

5.4.2 Chlorination with Chlorine Monofluoride

Chlorine monofluoride readily reacts both with organic and inorganic hydroxyl-containing compounds and some of their derivatives forming the respective hypochlorites according to the general scheme [71–76].

$$ROH + ClF \rightarrow ROCl + HF$$

Such reactions usually proceed vigorously, with heat evolution, and the end products—hypochlorites—are in most cases explosive. Therefore the reactions should be carried out with caution: gradual introduction of the reagent, the use of inert solvents, and strict control over the temperature are required. This procedure afforded high yields of chlorine nitrate [71,72], perfluoroalkyl hypochlorites [73], trifluoroacetyl hypochlorite [74], acyl hypochlorites [9,75], and chlorine triflate [76]. These compounds were used in subsequent syntheses as effective electrophilic reagents [9]. Acyl hypochlorites are also easily formed in the reaction of ClF with acyl anhydrides [75].

$$(CH_3CO)_2O + ClF \xrightarrow[0\,°C]{CCl_4} CH_3COOCl + CH_3COF$$

Chlorine monofluoride reacts with water very vigorously, forming a mixture of $ClFO_2$, Cl_2, O_2, and HF [77]. At low temperatures the main reaction products are Cl_2O and HF [71]. The organic and inorganic compounds with the N–H bond react with ClF in a similar way as the hydroxyl-containing compounds, forming the respective N-chloroamides [2,78].

$$RR'NH + ClF \rightarrow RR'NCl + HF$$

Chlorine monofluoride chlorinates organic compounds—benzene, chlorobenzene, and even nitrobenzene [79]. From chlorobenzene and toluene, *ortho*- and *para*-derivatives have been obtained in the ratio of 2:1, from nitrobenzene— *meta*-chloronitrobenzene. These are the examples of typical electrophilic aromatic substitution.

Table 2. Chlorofluorination of alkenes with ClF (1:1, mol)

N	Alkene		Products (%)			
			In $CHCl_3$ or $1,1,2\text{-}C_2Cl_3F_3$ (in HF[b])			
			Regioselectivity		Stereoselectivity	
		Yield	$C^1Cl\text{-}C^2F$	$C^1F\text{-}C^2Cl$	(% *anti*-adduct)	Ref.
1	$C^1H_2=C^2HCl$[c]	68	100	0		79
2	$CHCl=CHCl$[c]	75	88	12		14, 95
3	$C^1H_2=C^2Cl_2$[c]	70	100	0		95
4	$C^1HCl=C^2Cl_2$[c]	83				79, 95
5	$CCl_2=CCl_2$[c,d]	92				95
6	$C^1H_2=C^2HCH_2Cl$[c]	73	57	43		79
7	$C^1H_2=C^2HCH_2Br$[c,e]	84	55	45		95
8	$C^1H_2=C^2HCH_2OCOH$	48(38)	64(75)	36(25)		94
9	$C^1H_2=C^2HCH_2OCOOMe$	70(44)	60(68)	40(32)		94
10	$C^1H_2=C^2HCH_2OCOCCl_3$	82(f)	60	40		94
11	$Z\text{-}C^1HCl=C^2HCH_2Cl$[c]	95	2	98	90	94, 95
12	$E\text{-}C^1HCl=C^2HCH_2Cl$[c]	90	20	80	90	94, 95
13	$C^1H_2=C^2HCCl_3$[c]	96	0	100		94
14	$C^1F_2=C^2FCF_3$[c]	99	0	100		95, 96
15	$E\text{-}CF_3CF=CFCF(CF_3)_2$[c]	no reaction				95
16	$C^1H_2=C^2HCH_2COOMe$	89(60)	30(9)	70(91)		94

17	C^1H_2=C^2MeCOOMe	78(82)	53(95)	47(5)		94
18	C^1H_2=C^2MeCOCl	72	35	65		94
19	Z-C^1HCl=C^2FCOOMe[d]	71(70)	22(0)	78(100)	50(100)	94, 95
20	Z-C^1HF=C^2FCOOMe	[f](59)	(0)	(100)		95
21	C^1F_2=$C^2(CF_3)$COOMe	60(55)	0(0)	100(100)		95
22	Z-C^1ClMe=C^2HCOOEt	[f](72)	(0)	(100)	(50)	94
23	E-C^1ClMe=C^2HCOOEt	[f](72)	(0)	(100)	(50)	94
24	Z-C^1H(COOMe)=C^2HCOOMe	66(45)			100(100)	94
25	E-C^1H(COOMe)=C^2HCOOMe	68(48)			100(100)	94
26	Z-C^1H(COOMe)=C^2ClCOOMe	[d]50(86)	100(100)	0(0)	2:1(1:4)[g]	95
27	E-C^1H(COOMe)=C^2ClCOOMe	[d]50(80)	100(100)	0(0)	1:4(1:1)[g]	95
28	C^1Cl(COOMe)=C^2FCOOMe	60(70)	40(10)	60(90)	15:1(1:8)	95

[a] 15%(v/v) alkene

[b] 50%(v/v) alkene

[c] Because of insolubility of haloalkenes in HF, their chlorofluorination was not investigated

[d] ClF was passed through the solution of alkene in $C_2Cl_3F_3$ (2–10 ml of ClF per mol of alkene)

[e] Product of the bromine migration (25%) was also formed

[f] Complex mixture of unknown products

[g] Reaction was non-stereoselective. Configuration of diastereomers was not determined. One of them dominated in non-polar solvent, and another—in HF.

which may crystallise from the reaction mixture. Increased temperature of the reaction mixture may bring about an explosion [99].

The high selectivity of the substitutive fluorination is seen in the following example.

$$\text{CH}_2\text{BrCBrCH}_2\text{Cl} + \text{ClF} \xrightarrow[-\text{BrCl}]{} \left[\overset{+}{\text{CH}_2\text{BrCCH}_2\text{Cl}}\right] \longrightarrow \text{CH}_2\text{BrCFCH}_2\text{Cl}$$
$$\qquad\quad\;\;\underset{\text{Me}}{|} \qquad\qquad\qquad\underset{\text{Me}\;\;\text{F}^-}{|} \qquad\qquad\qquad\qquad\underset{\text{Me}}{|}$$

$$\text{CH}_2\text{BrCFCH}_2\text{Cl} + \text{ClF} \longrightarrow \left[\overset{+}{\text{CH}_2\text{CFCH}_2\text{Cl}}\right] \longrightarrow \left[\text{MeCH}_2\overset{+}{\text{CF}}\text{CH}_2\text{Cl}\right] \longrightarrow \text{MeCH}_2\text{CF}_2\text{CH}_2\text{Cl}$$
$$\qquad\quad\underset{\text{Me}}{|} \qquad\qquad\qquad\qquad\underset{\text{Me}\;\;\text{F}^-}{\diagdown|} \qquad\qquad\qquad\quad\underset{\text{F}^-}{} \qquad\qquad\qquad$$
$$\qquad\qquad\qquad\qquad\qquad\qquad\qquad\qquad\qquad\qquad\qquad\qquad 25 \qquad\qquad\qquad\qquad\qquad 26$$

Treatment of 1,2-dibromo-3-chloroisobutane with one mole of ClF leads to regioselective formulation of tertiary fluoride *10*; upon treatment with a second mole of ClF, the primary bromide elimination also proceeds easily and is accompanied by methyl migration. This results in the formation of a more stable α-fluorocarbocation *25* which adds the fluoride ion, giving the geminal difluoride *26*.

This example vividly shows gaseous ClF in non-polar solution to be a more powerful electrophilic agent than hexachloromelamine in HF. The latter is unable to eliminate primary bromine from haloalkane *10* (see Sect. 5.4.1). Substitution of bromine atoms in primary bromides proceeds in most cases with migration of vicinal substituents—methyl groups and chlorine atoms. Fluorination of 1-bromo-2-chloropropane (*27*) proceeds readily with chlorine migration. Hence the primary alkyl bromide *27* and its isomer *28* give the same fluorination product—1-chloro-2-fluoropropane *29*.

$$\text{MeCHClCH}_2\text{Br} \xrightarrow{\text{ClF}} \text{MeCHFCH}_2\text{Cl} \xleftarrow{\text{ClF}} \text{MeCHBrCH}_2\text{Cl}$$
$$\qquad 27 \qquad\qquad\qquad\qquad 29 \qquad\qquad\qquad\qquad 28$$

The driving force of all migrations in substitutive fluorinations is formation of a stable carbocation. The 1,2-hydrogen shift occurs even in non-polar media when a stable tertiary carbocation is formed.

$$\text{CH}_2\text{BrCHFMe} + \text{ClF} \longrightarrow \left[\overset{+}{\text{CH}_2}\text{CFMe}\right] \longrightarrow \left[\text{Me}\overset{+}{\text{C}}\text{FMe}\right] \longrightarrow \text{MeCF}_2\text{Me}$$
$$\qquad\qquad\qquad\qquad\qquad\qquad\qquad\quad\underset{\text{H}}{\diagdown} \qquad\qquad\underset{\text{F}^-}{} \qquad\qquad\qquad 100\%$$

$$\text{MeCHCH}_2\text{Br} + \text{ClF} \longrightarrow \left[\text{MeC}\overset{+}{\text{H}}\text{CH}_2\right] \left\{ \begin{array}{l} \longrightarrow \left[\text{t-Bu}^+\right] \longrightarrow \text{t-BuF} \\ \qquad\quad\;\underset{\text{F}^-}{} \qquad\quad 90\% \\ \\ \longrightarrow \left[\text{Me}\overset{+}{\text{C}}\text{HCH}_2\text{Me}\right] \longrightarrow \text{MeCHFCH}_2\text{Me} \end{array} \right.$$
$$\;\;\underset{\text{Me}}{|} \qquad\qquad\qquad\qquad\quad\underset{\text{Me}}{|} \qquad\qquad\qquad\qquad\qquad\qquad\qquad\underset{\text{F}^-}{}$$

Fluorination of propyl bromide in 1,1,2-$\text{C}_2\text{F}_3\text{Cl}_3$ at 0 to 10 °C also proceeds with hydrogen migration, but apart from 2-fluoropropane, it gives 1-fluoropropane.

$$CH_3CH_2CH_2Br + ClF \longrightarrow \begin{bmatrix} CH_3CH_2CH_2^+ \\ F^- \end{bmatrix} \longrightarrow \begin{bmatrix} CH_3\overset{+}{C}HCH_3 \\ F^- \end{bmatrix}$$

$$\downarrow \qquad\qquad\qquad \downarrow$$

$$CH_3CH_2CH_2F \qquad\qquad CH_3CHFCH_3$$
$$30\% \qquad\qquad\qquad 70\%$$

The reactions in non-polar solutions may lead to primary fluorides if the alternative hydrogen migration would lead to a carbocation that is only slightly more stable, or if the adjacent carbon is bonded with fluorine [98,99].

$$CH_2BrCH_2F + ClF \longrightarrow \begin{bmatrix} \overset{+}{C}H_2CH_2F \\ F^- \end{bmatrix} \longrightarrow \begin{cases} \overset{\longrightarrow}{/\!/} [CH_3CHF^+] \\ \longrightarrow CH_2FCH_2F \end{cases}$$

Comparison of the relative rates of fluorination of propyl and *iso*-propyl bromides showed the difference in their reactivities to be small (1:2) [99]. Nevertheless 1,2-dibromopropane reacts with one mole of ClF, selectively forming 1-bromo-2-fluoropropane.

$$CH_2BrCHBrCH_3 + ClF \begin{cases} \longrightarrow [CH_2Br\overset{+}{C}HCH_3] \xrightarrow{F^-} CH_2BrCHFCH_3 \\ \qquad\qquad 30 \\ \qquad\qquad \uparrow \\ \longrightarrow [\overset{+}{C}H_2CHBrCH_3] \\ \qquad\qquad 31 \end{cases}$$

Due to the liability of halogens (Cl, Br) to migration, the primary cation *31* formed by bromine elimination rearranges to a more stable secondary cation *30*, which accounts for the high selectivity of substitution.

Studies of the relative rates of substitutive fluorination of alkyl bromides in non-polar solution by the competing reactions method [99] have shown vicinal halogens to strongly slow down the fluorination. For secondary bromides, the following reactivity sequence is observed.

$$MeCHBrMe \gg MeCHBrCH_2Cl \gg CH_2ClCHBrCH_2Cl$$

The same is observed for primary bromides.

$$MeCH_2CH_2Br \gg MeCHClCH_2Br$$

The geminal halogens slow down fluorination only slightly. Thus ethyl bromide and 1-bromo-1-chloroethane are fluorinated at about equal rates, and in the series of dibromoethanes the ratio of fluorination rates is the following.

$$CH_2BrCH_2Br : CH_2BrCHClBr : CH_2BrCCl_2Br = 10 : 3 : 1$$

In this case, with the stoichiometric amount of ClF, only α-halogenfluorides are formed. Such a high selectivity obviously results from the fact that, e.g., in appropriate substrate or its solution in inert solvent at -50 to $-10\,°C$. In this way, *tert*-butyl chloride, 1,2-dichloropropane, and 1,2,3-trichloropropane are easily fluorinated, giving nearly quantitative yields. Even chloroform is fluorinated in these conditions to a high degree.

$$CH_3CHClCH_2Cl + ClF \xrightarrow{C_2F_3Cl_3} CH_3CHFCH_2Cl$$

$$CH_2ClCHClCH_2Cl + ClF \rightarrow CH_2ClCHFCH_2Cl$$

$$CHCl_3 + ClF \rightarrow CHFCl_2$$

The above examples show the high selectivity of fluorination: the primary chlorine atoms remain intact. For deeper fluorination, activation of ClF by HF or Lewis acids is necessary [102].

Chlorine monofluoride in anhydrous HF substitutes chlorine in organic chloro-derivatives by fluorine at -60 to $10\,°C$. The reaction proceeds readily when gaseous ClF is passed into the HF solution or stirred emulsion of a chloro-derivative, according to the scheme.

$$R\!-\!Cl \cdots Cl\!-\!F \cdots H\!-\!F \longrightarrow R^+ + Cl_2 + HF_2^-$$
$$R^+ + HF_2^- \longrightarrow RF + HF$$

This reaction has the same limitation as the substitution of bromine by fluorine: the chlorine atom of the CHCl group in α-position to the alkoxycarbonyl group is not substituted. When one mole of ClF reacts with one mole of methyl 2,3-dichloropropanoate, only the β-chlorine atom is substituted.

$$CH_2ClCHClCOOMe + ClF \xrightarrow{HF} CH_2FCHClCOOMe$$

$$\xrightarrow{ClF/HF} \nrightarrow CH_2FCHFCOOMe$$

Upon passing of a second mole of ClF, the α-chlorine atom remains intact, and the electrophilic attack of ClF is directed at the carbonyl oxygen (see Sect. 5.4.5). Introduction of carbocation-stabilising substituents into the α-position leads to substitution of α-chlorine by fluorine.

$$CH_2ClCCl_2COOMe + ClF \xrightarrow{HF} CH_2ClCFClCOOMe$$

$$\xrightarrow[HF]{ClF} CH_2ClCF_2COOMe$$

$$\underset{\underset{Me}{|}}{CHClFCClCOOMe} + ClF \xrightarrow{HF} \underset{\underset{Me}{|}}{CHF_2CClCOOMe}$$

$$\xrightarrow{ClF} \underset{\underset{Me}{|}}{CHF_2CFCOOMe}$$

The HF solution promotes hydrogen migrations with formation of more stable carbocations in substitutive fluorination. In the reaction of 1,2,3-trichloropropane with one mole of ClF, the central chlorine atom is selectively substituted, and further attack on the primary chlorine atom of 1,3-dichloro-2-fluoropropane is accompanied by hydrogen migration. This leads to 1-chloro-2,2-difluoropropane, in which the chlorine atom remains intact in these conditions.

$$CH_2ClCHClCH_2Cl + ClF \xrightarrow{HF} CH_2ClCHFCH_2Cl \xrightarrow[HF]{ClF} \left[\overset{+}{C}H_2CFCH_2Cl \underset{H \quad F^-}{\big|} \right] \longrightarrow CH_3CF_2CH_2Cl$$

$$\xrightarrow[HF]{ClF} \not\rightarrow CH_3CF_2CH_2F$$

The catalytic amount of antimony halide (3–5%) enhances the fluorinating ability of chlorine monofluoride in a non-polar solution [102]. This may be explained by the formation of fluorodichloronium hexafluoroantimonate [28], in which the electrophilic activity of chlorine is so high that such complex can fluorinate 1,1,2-trichlorotrifluoroethane to 1,2-dichlorotetrafluoroethane at -40 to $-60\,°C$.

$$CClF_2CCl_2F + Cl_2F^+SbF_6^- \xrightarrow[-Cl_2, -ClF]{} [CClF_2\overset{+}{C}ClF] \quad SbF_6^-$$

$$\xrightarrow[-SbF_5]{} CF_2ClCF_2Cl$$

The antimony salt-catalysed substitutive fluorination by chlorine monofluoride proceeds less selectively than the non-catalysed or ClF–HF-catalysed ones. Thus the reaction of 1,2,3-trichloropropane with an equivalent amount of ClF in the presence of 5% of SbCl$_5$ gave, apart from the main product, 1,3-dichloro-2-fluoropropane *41*, substantial amounts of 3-chloro-1,2-difluoropropane *14*, 3-chloro-1,1-difluoropropane *42*, and 1,1,3-trifluoropropane *43*.

$$CH_2ClCHClCH_2Cl + ClF \xrightarrow{(SbCl_5)} \left[\underset{Cl}{\overset{Cl}{CH_2}} \overset{+}{C}H - CH_2 \; SbF_6^- \right] \longrightarrow \left[\overset{+}{C}HClCH_2CH_2Cl \; SbF_6^- \right]$$

$$\left[CH_2FCHClCH_2Cl \right] + CH_2ClCHFCH_2Cl \qquad \left[CHFClCH_2CH_2Cl \right]$$
$$41 \qquad\qquad CHF_2CH_2CH_2Cl$$
$$CHF_2CHFCH_2Cl \qquad\qquad 42$$
$$14 \qquad\qquad CHF_2CH_2CH_2F$$
$$43$$

Thus fluorination with chlorine monofluoride possesses wide possibilities for the selective synthesis of various mono- and polyfluorides, with variation of the reaction temperature and catalysing additions.

It is interesting to note that in the reaction of dithiane 52 with chlorine monofluoride, upon gradual increase of the temperature from -196 to $0\,°C$, the chlorofluorinated derivative 53 is formed initially, which is further converted to the perfluorinated product. Both steps give quantitative yields [111].

Perfluoroalkylsulfenyl chlorides and disulfides also readily undergo the oxidative fluorination by chlorine monofluoride, forming trans-R_FSF_4Cl [112].

$$R_FSCl \atop R_FSSR'_F \quad \xrightarrow{\quad ClF \quad} R_FSF_4Cl$$

R_F, R'_F = perfluoroalkyl

The perfluoroalkyl derivatives $R_FSF_4R'_F$ and R_FSF_4Cl are chemically inert, whereas the fluorosulfuranes may be used in organic synthesis. Thus the reaction of $R_FSF_2R'_F$ with anhydrous HCl in a tube, in the presence of glass powder, gives perfluoroalkyl sulfoxides, which in their turn may undergo oxidative fluorination by chlorine monofluoride.

$$R_FSF_2R'_F \xrightarrow[2)\,H_2O]{1)\,HCl} R_FSOR'_F$$

$$R_FSOR'_F + ClF \xrightarrow[-78\,°C,\,3\,h]{} R_FSOF_2R'_F$$

R_F, R'_F = CF_3, C_2F_5

In this way perfluoroalkylsulfuryl difluorides were first obtained [113].

   The reactions of chlorine monofluoride with some bis(perfluoroalkyl) selenides and bis(perfluoroalkyl) diselenides have been studied in detail in one of recent works [114]. The oxidative fluorination of perfluoroalkyl selenides to $(R_F)_2SeF_2$ smoothly proceeds in the conditions described above for the reactions with analogous sulfides. The reactions give very high yields of products.

$$(R_F)_2Se + 2ClF \xrightarrow[-78\,°C]{} (R_F)_2SeF_2 + Cl_2$$
$$98\text{--}100\%$$

$$R_F = CF_3, C_2F_5$$

But the Se(IV) derivatives produced in these reactions do not react with ClF, as opposed to the sulfur-containing analogues, even after keeping the reaction mixture at room temperature for a long time. Compounds $(R_F)_2SeF_4$ are not formed in more rigid conditions either.

The reactions of chlorine monofluoride with bis(perfluoroalkyl) diselenides also have some peculiarities. The reaction of bis(perfluoroethyl) diselenide with ClF in the mole ratio of $1:6$ at $-78\,°C$ gives perfluoroethylselenium trifluoride 54 with a quantitative yield.

$$(C_2F_5)_2Se + 6ClF \xrightarrow[-78\,°C]{} 2C_2F_5SeF_3 + 3Cl_2$$

Compound 54 at room temperature consumes one more mole of ClF during several hours, forming *trans*-perfluoroethylselenium chlorotetrafluoride 55.

$$C_2F_5SeF_3 + ClF \xrightarrow[20\,°C]{} C_2F_5SeF_4Cl$$
$$54 \qquad\qquad 55 \quad 40\%$$

By contrast with the chemically inert derivatives of S(VI), compound 55 is itself a vigorous oxidant, as indicated by the reaction of this compound with metallic mercury, in which HgF_2 and $HgCl_2$ are formed [114].

The oxidative fluorination of bis(trifluoromethyl) telluride by chlorine monofluoride leads to bis(trifluoromethyl)tellurium difluoride, which is a white, easy-hydrolysable substance [115].

5.5 Fluorination with Halogen Polyfluorides

Section 2 considered in detail methods of synthesis and properties of halogen polyfluorides. Among these, the most intensively studied are ClF_3, BrF_3, and IF_5. Being liquids, these compounds are preferable to gaseous chlorine monofluoride, as they do not require any special equipment. They may be stored in copper, steel, nickel or Teflon vessels and used when required. The most aggressive of all halogen fluorides—chlorine trifluoride—has not found use in controlled fluoroorganic synthesis. Bromine trifluoride and iodine pentafluoride as far less aggressive and dangerous reagents attracted greater attention, so this Section discusses the results of their preparative uses and recent mechanistic studies of their reactions with organic substrates.

$$CF_2=CFCF_3+1/3(BrF_3+Br_2) \xrightarrow[20\,°C]{} (CF_3)_2CFBr$$
$$46\%$$

The reaction of perfluoroalkenyl ethyl ethers with the BrF_3–Br_2 system proceeds both with the addition of BrF at the multiple bond and substitution of α-hydrogen in the ethyl group by fluorine.

$$(CF_3)_2C=CFOCH_2CH_3+BrF_3+Br_2 \xrightarrow[60\,°C]{} (CF_3)_2CBrCF_2OCHFCH_3$$
$$28\%$$

It had long been thought impossible to carry out the selective fluorination of polyhydro-substrates by halogen polyfluorides: as a rule, these reactions gave a complex mixture of the products of fluorination and chlorination (bromination) of different degrees [5,6,124,125].

$$CHCl=CHCl+ClF_3 \xrightarrow[55\,°C]{} C_2H_2Cl_3F+C_2H_2Cl_2F_2$$
$$24\% \qquad\qquad 40\%$$

$$\begin{array}{c}CHCl=CCl\\ |\\ CHCl=CCl\end{array}+ClF_3 \xrightarrow[20\,to\,150\,°C]{} C_4F_6Cl_4+C_4F_5Cl_5+C_4F_4Cl_6+C_4F_3Cl_7$$
$$19\% \qquad\quad 45\%$$

The mixture of octafluorotetrabromocyclohexane isomers $C_6F_8Br_4$ was obtained by treatment of hexabromobenzene with ClF_3, IF_5 or IF_7 in bromine at −10 to 60 °C [126]. One of the first examples of selective fluorination of unsaturated polyhydro-substrates was the reaction of bromine trifluoride with lower nitriles and acetone in anhydrous HF [127].

$$RCN+BrF_3 \xrightarrow[-20\,to\,15\,°C]{HF} RCF_3$$

$$R = Me,\ Et,\ CH_2Cl$$

$$CH_3COCH_3+BrF_3 \xrightarrow{HF} CH_3CF_2CH_3$$

These reactions possibly proceed via the intermediate addition at the $C\equiv N$ and $C=O$ bonds. However in the case of ethyl methyl ketone, there occurs the predominant fragmentation of the molecule at the C–C bonds. The conditions have been found, in which BrF_3 without molecular bromine addition smoothly reacts with mono- and dihaloalkenes with a short carbon chain, giving the equimolar mixture of bromofluorides and difluorides in a high total yield [128]. The reaction is carried out by gradually adding liquid BrF_3 (0.5 mole per one multiple bond) to the diluted (3 to 5% v/v) alkene solution in $C_2F_3Cl_3$.

$$CHBr=CHBr+1/2\,BrF_3 \xrightarrow[10\,to\,20\,°C]{C_2F_3Cl_3} CHBr_2CHBrF+CHBr_2CHF_2$$
$$38\% \qquad\qquad 45\%$$

$$CH_2=CClCH_2Cl + 1/2\ BrF_3 \rightarrow CH_2BrCClFCH_2Cl + CH_2ClCF_2CH_2Cl$$
$$35\%33\%$$

The equimolar BrF_3–Br_2 mixture may be used for bromofluorination of multiple bonds of α,β,-unsaturated esters [128].

$$CH_2=CMeCOOMe + 1/3(BrF_3 + Br_2) \xrightarrow[\text{10 to 20 °C}]{C_2F_3Cl_3} CH_2BrCFMeCOOMe$$
$$46\%$$

$$+ CH_2FCBrMeCOOMe$$
$$36\%$$

$$CH_2=CClCOOMe + 1/3(BrF_3 + Br_2) \rightarrow CH_2BrCClFCOOMe$$
$$42\%$$

The reaction of unsaturated esters with pure BrF_3 fails to give satisfactory yields of the bromofluoro- and difluoro-adducts.

Detailed studies have been carried out on the products of the reaction of polyfluoroaromatic compounds with the BrF_3–Br_2 system [129,130], pure BrF_3 [131], and polyfluorohalogenonium salts [131–133].

In the reactions of hexafluorobenzene, pentafluorobenzene, bromopenta-fluorobenzene, octafluorotoluene, decafluoro-p-xylene, 2,3,4,5,6-pentafluoro-toluene, and 2,3,4,5,6-pentafluoroanisole with an equimolar BrF_3–Br_2 mixture (in the mole ratio of substrate: $BrF_3 : Br_2 = 1 : 1 : 1$, i.e. 3 equivalents of "BrF" per 1 mole of polyfluoroarene) in inert solvents at $0\,°C$, at first there occurs fluorination of the aromatic ring with formation of substituted polyfluoro-1,4-cyclohexadienes, which further add BrF at one or two multiple bonds, giving bromopolyfluorocyclohexenes and -cyclohexanes (the products were isolated after decomposition of the BrF_3 excess with water).

If the starting perfluoroarene has an electron-accepting substituent (except fluorine atoms), for example, the CF_3 group, then in the diene product it occupies

The exchange of bromine and chlorine for fluorine in haloalkanes in the reaction with IF_5 proceeds at high temperatures with difficulty, and has no preparative value [7]. The interaction of chloro- and bromo-deivatives with bromine trifluoride usually led to the formation of a mixture of products of various fluorination degrees [1,6]. The reaction of BrF_3 with α,α,α'-trichlorohexa-fluoropiperidine at $-50\,°C$, however, gave one product—N-bromoperfluoro-piperidine [134].

An efficient reagent for the substitutive fluorination of bromopolyfluoro-alkanes is the BrF_3–Br_2 system [135]. The reaction is carried out by gradually adding bromofluoroalkane to a mixture of BrF_3 (1/3 mole per one g-atom of bromine) and molecular bromine at 0 to $50\,°C$, with simultaneously distilling off the low-boiling fluorination product.

$$CF_3CHBr_2 + BrF_3 + Br_2 \rightarrow CF_3CHFBr + CF_3CHF_2$$
$$ 92\% 5\%$$

$$CF_2BrCHBr_2 + BrF_3 + Br_2 \rightarrow CF_3CHBrF$$
$$ 77\%$$

$$CF_2BrCBr_3 + BrF_3 + Br_2 \rightarrow CF_2BrCFBr_2 + CF_2BrCF_2Br$$
$$ 79\% 4\%$$

This procedure may be used to fluorinate lower bromofluoroalkanes where the C:H ratio is not higher than 1:1, at a small time of contact of substrate with reagent, as the reaction is carried out with a large excess of BrF_3 relative to substrate in the beginning and in the process of the reaction. Molecular bromine as a solvent surely plays an essential role in these reactions, promoting ionisation of the BrF_3 molecule, as indicated by the relatively high electric conductivity of the BrF_3–Br_2 system [1]. In addition, liquid bromine as medium in the reactions involving the carbocation intermediates is similar in its action to such a well-investigated electrophilic solvent as H_2SO_4 [136]. In more recent works it has been shown that pure BrF_3 may be used as an efficient agent of selective fluorodebromination of bromo-derivatives (with perfluorocarbons as solvent, 10 to $25\,°C$ [137,138]. The reaction follows the scheme.

$$RBr + 1/3BrF_3 \rightarrow RF + 2/3Br_2$$

The reaction is carried out by gradually adding bromine trifluoride to a solution of the bromo-derivative (5 to 30% v/v), with vigorous stirring. The higher is the content of halogens in a substrate, the more concentrated solution may be used in the synthesis. In the process of fluorination, all the three fluorines of the BrF_3 molecule at the given stoichiometry are effectively used. The chlorine atoms of the molecule usually remain intact. The fluorodebromination of bromoalkanes by bromine trifluoride is non-stereospecific and proceeds with carbocationic rearrangements considered in Sect. 5.4.4. The typical examples of substitutive fluorination by BrF_3:

$$CH_2ClCH_2Br + BrF_3 \rightarrow CH_2FCH_2Cl$$
$$ 80\%$$

$$CH_2BrCHBrCH_2Cl + BrF_3 \rightarrow CH_2FCHFCH_2Cl$$
$$85\%$$

$$CH_2BrCFMeCH_2Cl + BrF_3 \rightarrow CH_2FCFMeCH_2Cl + MeCH_2CF_2CH_2Cl$$
$$6\% \qquad\qquad\qquad 67\%$$

$$CH_3(CH_2)_3CHBrCH_2Br + BrF_3 \rightarrow CH_3(CH_2)_3CHFCH_2Br$$
$$72\%$$

pref-CHClBrCHFCl ⎤
                  ⎥ $\xrightarrow{\text{BrF}_3}$ pref-CHClFCHFCl
parf-CHClBrCHFCl ⎦
$$70\text{–}80\%$$

In the reactions under discussion, bromine trifluoride is much less reactive than chlorine monofluoride. Thus ClF readily and completely reacts with 1-bromo-2,3-dichloropropane at -30 to $-40\,°C$, whereas with BrF_3 at the same temperature there is practically no reaction [137].

$$CH_2ClCHClCH_2Br \quad \begin{array}{l} \xrightarrow[-40\,°C]{\text{ClF}} CH_2ClCHFCH_2Cl \\ \xrightarrow[-40\,°C]{\text{BrF}_3} /\!/ \\ \xrightarrow[20\,°C]{\text{BrF}_3} CH_2ClCHFCH_2Cl \end{array}$$

The difference in the reactivities of BrF_3 and ClF is still more vividly seen in the reactions with bromine-containing esters. Thus the reaction of methyl dibromo-isobutyrate 72 with ClF proceeds to an end when the stoichiometric amount of ClF is passed through this solution for 1 h, whereas the reaction with BrF_3 takes about 30 h.

$$CH_2BrCBrMeCOOMe + 1/3\,BrF_3 \xrightarrow[20\,°C,\ 30h]{} CH_2BrCFMeCOOMe$$
$$72 \qquad\qquad\qquad\qquad\qquad\qquad\qquad ·55\text{–}60\%$$

$$+ CH_2FCHBrMeCOOMe$$
$$15\text{–}20\%$$

In this case, such a low reactivity of BrF_3 is also explained by the fact that the COOMe group deactivates BrF_3 in the substitutive fluorination reaction. Fluorination of methyl 2,3-dibromo-2-chloropropanoate 73 in similar conditions proceeds much faster, which may be explained by the decreased basicity of the COOMe group of ester 73 [138].

$$CH_2BrCBrClCOOMe + BrF_3 \xrightarrow[20\,°C,\ 10h]{} CH_2BrCFClCOOMe$$
$$73 \qquad\qquad\qquad\qquad\qquad\qquad 74$$

Of course, this is not due to the fact that chlorine at the reaction centre promotes substitution more than the CH_3 group in ester 72, as in the reaction of BrF_3 will a mixture of esters 72 and 73, compounds 72 is fluorinated first.

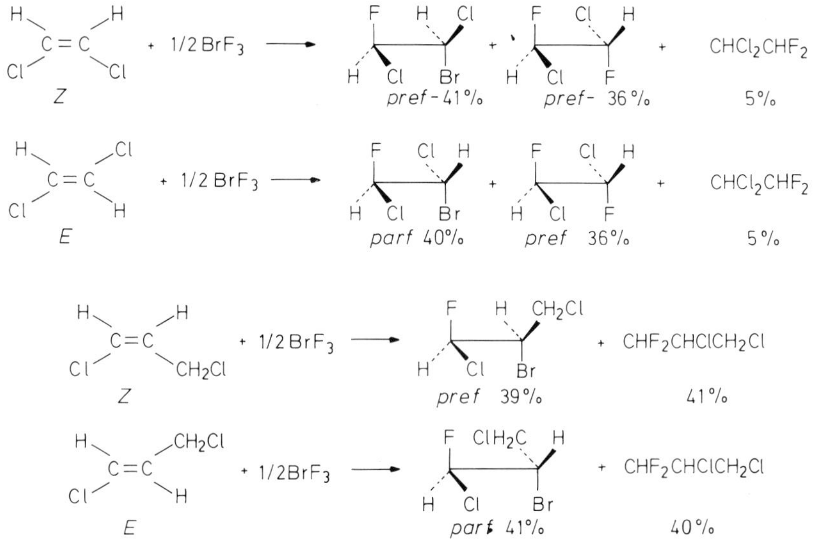

On the basis of these data, the reactions of fluorination and bromofluorination of alkenes by bromine trifluoride were suggested [128] to proceed via different independent transition states or intermediates (Scheme 1).

Scheme 1

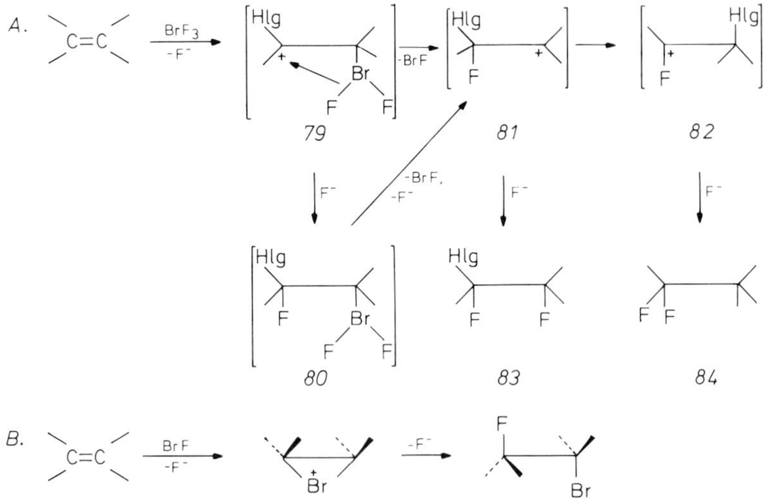

The electrophilic attack at the olefin double bond in non-polar medium is effected by the BrF_3 molecule. The bromine (III)-containing carbocation 79 formed in this process eliminates the BrF molecule to form the β-fluorocarbocation 81. The latter is stabilised by capturing the fluoride ion, forming vicinal difluoride *83*, or is rearranged to a more stable α-fluorocarbocation *82*, which is further transformed to the more geminal difluoride *84*. Carbocation *79* possibly manages to capture the fluoride ion, giving an unstable intermediate *80*, which further also decomposes to bromine monofluoride and carbocation *81*. A relatively stable adduct of type *80* with trivalent iodine was obtained in the addition reaction of iodine tris(fluorosulfate) with perfluoroolefins [141].

Thus the initial stage (stage *A*) of the reaction involves the addition of two fluorine atoms at the C=C bond, and the C=C bond is bromofluorinated by bromine monofluoride liberated at the first stage, according to the ordinary scheme of electrophilic *anti*-addition (stage *B*) (cf. Sect. 5.3.1).

The suggested reaction mechanism adequately accounts for the unfavourable (from the first sight) migration of vinyl halogen (Cl, Br) leading to geminal difluorides *84*, and the sharp difference in the stereochemistry of formation of vicinal difluoro- and bromofluoro-adducts. As mentioned in Sect. 5.5.2, the reaction of perfluoroaromatic compounds with BrF_3 initially gives the products of fluorination of the aromatic ring—perfluorocyclohexadienes, and then—their adducts with BrF [129]. Bastock and his co-workers [129] carried out the semi-empirical thermodynamic calculations of the reactions of BrF_3–Br_2 with per-fluoroarenes considering ionisation potentials of fluoroaromatic compounds, and suggested a general scheme of one-electron oxidation of these substrates to radical cations. This opinion is shared by the scientists who studied the reactions of fluorohalogenonium cations (ClF_2^+, BrF_2^+, BrF_4^+, and IF_4^+) with perfluoroaro-matics [131–133].

All unmarked bonds to fluorine

Radical cation *85* was recorded by ESR spectroscopy in the reaction of perfluoronaphthalene with ClF_3, BrF_3, BrF_5, and XeF_2 at $-50\,^\circ C$ [133].

$$C_{10}F_8 + BrF_3 \xrightarrow[-50\,^\circ C]{SO_2FCl} C_{10}F_8^{+\cdot}$$

Fluorohalogenonium cations XF_n^+ (X = Cl, Br, I) have a higher electron affinity than the starting halogen fluorides, being thus more reactive. For example, $EA(BrF_2^+)$ and $EA(ClF_2^+)$ are equal to 11–12 eV. The ionisation potential of most perfluoroarenes ranges from 9 eV (octafluoronaphthalene) to 10.6 eV (nitro-pentafluorobenzene).

temperature, then poured onto ice. The organic layer was washed with water, then with the sodium sulfite solution, dried with $MgSO_4$ and distilled to give 22 g (59%) of haloalkane *10*: b.p. 55 °C (13 mm), d_4^{20} 1.5994, n_D^{20} 1.4630.

4. 1*R**,2*R**-1-fluoro-1,2,3-trichloropropane *23* [94]

Into a solution of 134 g (1.2 mol) of *Z*-1,3-dichloropropene in 500 ml of chloroform, 30 l of gaseous ClF was passed for 60 to 70 min at 20 to 30 °C. After the reaction was completed, chlorine monofluoride was not absorbed (KI probe at the outlet of the reactor). The product was distilled to give 184 g (92%) of fluorotrichloropropane *23*: b.p. 143 °C, d_4^{20} 1.4818, n_D^{20} 1.4554.

5. Dimethyl *threo*-2-chloro-3-fluorosuccinate *24* [94]

Into a solution of 85 g (0.71 mol) of dimethyl maleate in 350 ml of chloroform, 0.8 mol of ClF was passed by the above-described method. The product was distilled to give 77.3 g (66%) of *threo*-diastereomer *24*: b.p. 88 to 90 °C (2 mm), d_4^{20} 1.3645, n_D^{20} 1.4357.

6. 1,2,3-Trifluoropropane *45* [98]

A glass cylinder reactor with a reflux condenser, thermometer, and polyethylene bubbler was charged with 281 g (1 mol) of 1,2,3-tribromopropane, and ClF was passed at 0 to -5 °C and at 25 l h^{-1}, for 3 h. The reaction mixture was treated with NaOH solution under cooling to disappearance of bromine colouring, then dried over $MgSO_4$ and distilled to give 69 g (71%) of trifluoropropane *45*: b.p. 70 °C, d_4^{20} 1.2680, n_D^{20} 1.3250.

7. Methyl 2,3-difluoroisobutyrate *46* [98,102]

A. Gaseous ClF (50 l) was passed through 257 g (1 mol) of methyl 2,3-dibromo-isobutyrate during 2 h, as described above. Then chloroform (200 ml) was added to the reaction mixture, and the mixture was washed with Na_2SO_3 solution to give 125 g (93%) of ester *46*: b.p. 49 °C (18 mm), d_4^{20} 1.1877, n_D^{20} 1.3850.

B. A polyethylene cylinder reactor was charged with a solution of 86 g (0.5 mol) of methyl 2,3-dichloroisobutyrate in 60 ml of anhydrous HF, and 26 l of ClF was passed at -20 to -40 °C for 2 h. The reaction mixture was poured onto ice, the organic layer was separated, washed with $NaHSO_3$ solution, dried over $MgSO_4$ and distilled to give 57 g (83%) of ester *46*: b.p. 50 °C (18 mm), n_D^{20} 1.3850.

8. 1,3-Dichloro-2-fluoropropane *41* [102]

In a glass cylinder reactor were placed 147.5 g (1 mol) of 1,2,3-trichloropropane and 50 ml of 1,1,2-trichlorotrifluoroethane, and ClF was passed at -25 to -30 °C for 3 h (10 l h^{-1}). The reaction mixture was washed with Na_2SO_3 solution, dried and distilled to give 111 g (84%) of haloalkane *41*: b.p. 127 °C, d_4^{20} 1.4350, n_D^{20} 1.4353.

9. 1-Chloro-2,2-difluoro-3-oxabutane *50* [106]

A polyethylene cylinder reactor was charged with 27 g (0.25 mol) of methyl chloroacetate and 20 ml of anhydrous HF and cooled to -60 to -40 °C. Chlorine monofluoride (15 l, 0.6 mol) was passed into the solution for 25–30 min. The reaction mixture was poured onto ice. The organic layer was washed with ice water, dried and distilled to give 26 g (80%) of ether *50*: b.p. 76 °C, d_4^{20} 1.2850, n_D^{20} 1.3530.

10. 1-Chloro-2,2,4-trifluoro-3-oxapentane *51* [106]

Into a solution of 24.5 g (0.2 mol) of ethyl chloroacetate in 20 ml of HF, 18 l (0.7 mol) of ClF was passed under the above conditions, for 40–50 min. The procedure gave 19.6 g (61%) of ether *51*: b.p. 47 °C (115 mm), d_4^{20} 1.3045, n_D^{20} 1.3530.

11. *Pref*-2-bromo-1-fluoro-1,3-dichloropropane *39* and 2,3-dichloro-1,1-difluoropropane *40* [128]

In a Teflon reactor with a nickel stirrer, nickel réflux condenser and quartz dropping funnel, were placed 24.4 g (0.22 mol) of Z-1,3-dichloropropene and 380 ml of 1,1,2-$C_2F_3Cl_3$. With stirring and cooling the reactor with ice water 15.2 g (0.11 mol) of BrF_3 was gradually added for 30–40 min at 5 to 10 °C. The reaction mixture was washed with aqueous sodium sulfite, dried with $CaCl_2$, and distilled on a column (5 th.pl.). This procedure gave 13.3 g (40%) of halopropane *40*: b.p. 55 °C (100 mm), d_4^{20} 1.4460, n_D^{20} 1.4040. Then 19.1 g (41%) of *pref*-diastereomer *39* was isolated: b.p. 40 °C (5 mm), d_4^{20} 1.8233, n_D^{20} 1.4908.

12. 4-Bromononafluorocyclohexene *71* [129]

A nickel reactor with a stirrer, dropping funnel, and copper reflux condenser was charged with 10 g (0.054 mol) of hexafluorobenzene, 40 ml of perfluoromethyl-cyclohexane (solvent), and 8.6 g (0.054 mol) bromine. Then 7.3 (0.054 mol) of BrF_3 was slowly added with stirring at 0 °C. The mixture was stirred for 1 h at 0 °C, whereupon water was added to decompose unchanged BrF_3. The organic layer was washed with $NaHSO_3$ solution and dried over $MgSO_4$. Distillation of the reaction mixture gave 7.6 g (43%) of cyclohexene *71*: b.p. 99 °C, and 1.1 g (9%) of octafluorocyclohexa-1,4-diene.

13. 2-Bromo-1,1,1,2-tetrafluoroethane *78* [135]

A Monel flask provided with a nickel condenser, nickel dropping funnel and nickel stirrer was charged with 54 g (0.4 mol) of BrF_3 and 236 g of bromine. To this mixture, 247 g (1 mol) of 2,2-dibromo-1,1,1-trifluoroethane was added for 15 min, with cooling of the flask to 25 to 35 °C and stirring. The gaseous products were simultaneously collected in a glass trap cooled to −78 °C. Then the reaction mixture was heated to 50 °C to complete the reaction. The reaction products were washed with the cooled NaOH solution to remove bromine. This procedure gave 156 g (85%) of haloethane *78*: b.p. 8 to 9 °C (with a 0.5% admixture of CF_3CHF_2).

14. 1-Bromo-3-chloro-2-fluoropropane *12* and 3-chloro-1,2-difluoropropane *14* [138]

A Teflon reactor with a nickel stirrer, nickel reflux condenser, thermometer in a nickel pocket and quartz dropping funnel was charged with a solution of 170 g (0.67 mol) of 1,2-dibromo-3-chloropropane in 400 ml of 1,1,2-$C_2F_3Cl_3$. Then 50.4 g (0.37 mol) of BrF_3 was added dropwise, with vigorous stirring for 3 h at 5 to 10 °C. The reaction mixture was stirred for 1 h more, then treated with the $NaHSO_3$ solution, dried and distilled to give 24.7 g (25%) of halopropane *14*, b.p. 96 °C (750 mm) and 69.5 g (60%) of halopropane *12*, b.p. 125 to 130 °C (100 mm), d_4^{20} 1.7408, n_D^{20} 1.4700.

147. Zefirov NS, Kozmin AS, Sorokin VD (1984) J. Org. Chem. 49:4086
148. Zefirov NS, Kozmin AS, Sorokin VD, Zhdankin VV (1986) Zh. Org. Khim. 22:898
149. Beringer FM, Schultz HS (1955) J. Amer. Chem. Soc. 77:5533
150. Schmid GH, Gordon JW (1983) J. Org. Chem. 48:4010
151. Svetlakov NV, Moisak IE, Averko-Antonovich IG (1969) Zh. Org. Khim. 5:2105
152. Svetlakov NV, Moisak IE, Shafigulin NK (1971) Zh. Org. Khim. 7:1097
153. MacDonald TL, Narasimhan N, Burka LT (1980) J. Amer. Chem. Soc. 102:7760
154. Davidson RI, Kropp PJ (1982) J. Org. Chem. 47:1904
155. Schack CJ, Christe KO (1980) J. Fluor. Chem. 16:63
156. Fokin AV, Studnev YN, Rapkin AI, Tatarinov AS, Seryanov YV: Izv. Akad. Nauk SSSR. Ser. Khim. 1985:1635
157. Fokin AV, Studnev YN, Rapkin AI, Chelikin VG, Verenikin OV: Izv. Akad. Nauk SSSR. Ser. Khim. 1985:659
158. Katsuhara Y, DesMarteau DD (1980) J. Amer. Chem. Soc. 102:2681
159. Schack CJ, Pilipovich D, Christe KO (1975) Inorg. Chem. 14:145
160. Bach RD, Holubka JW, Taaffee TH (1979) J. Org. Chem. 44:35
161. Brit. Pat 805764 (1953); (1953) Chem. Abs. 53:10035
162. Fokin AV, Rapkin AI, Seryanov YV, Studnev YN: Izv. Akad. Nauk SSSR. Ser. Khim. 1986:2734
163. Arapov OV, Rudenko AP, Zarubin MY (1985) Zh. Org. Khim. 21:168
164. Fokin AV, Semin TK, Raevskii AM, Gyshchin SI, Rapkin AI, Tatarinov AS, Studnev YN: Izv. Akad. Nauk SSSR. Ser. Khim. 1986:244
165. Zefirov NS, Makhonkov DI (1982) Chem. Rev. 82:615
166. Barter RM, Littler JS: J. Chem. Soc. (B) 1967:205

6 New Uses of Sulfur Tetrafluoride in Organic Synthesis

Anatolii Ivanovich Burmakov, Boris Vasil'evich Kunshenko, Lyubov' Antonovna Alekseeva and Lev Moiseevich Yagupolskii

Polytechnic Institute, Shevchenko St. 1, 270044, Odessa, USSR and Institute of Organic Chemistry, Murmanskaya St. 5, 252660, Kiev, USSR

Contents

6.1   Introduction

Since the publication in 1959–1960 of the pioneering works [1,2] which showed that by using sulfur tetrafluoride it is possible to effect the selective substitution of carbonyl oxygen in aldehydes, ketones, carboxylic acids, and of the hydroxy group in alcohols by fluorine, interest in this fluorinating agent has been constantly growing. Similar reactions with sulfur tetrafluoride have been reported for the compounds in which oxygen is double-bonded with phosphorus, arsenic [3,4], germanium [5], antimony [6], and iodine [7,8], as well as for dicarbonyl compounds [9]. The reactions of SF_4 with compounds containing double and triple carbon–nitrogen bonds gave the representatives of a new class of compounds–iminosulfur difluoride derivatives [10, 11].

Several reviews on the reactions of organic compounds with SF_4 have been published [12–15], but they cover literature only until 1973. Recently a lot of new data considerably expanding our knowledge on the synthetic possibilities of SF_4 have appeared. This was also stimulated by the development of convenient methods for its synthesis [16–22].

Dialkylaminosulfur trifluorides formed in the reaction of SF_4 with silylated amines were found to be the SF_4 analogues capable of substituting the hydroxy groups and carbonyl oxygen by fluorine [23,24].

The reactions of SF_4 with polyfunctional organic compounds have been studied, and methods have been developed for the synthesis of aliphatic, aromatic, and heterocyclic fluorine-containing alcohols, ketones, carboxylic acids, etc.

The efficiency of SF_4 as an agent for the substitution of oxygen by fluorine was found to sharply increase when the reactions of oxygen-containing organic compounds with SF_4 are carried out in anhydrous hydrogen fluoride.

More and more data are appearing indicating that in a number of reactions SF_4 may have an oxidating activity. The use of fluorinating systems on the basis of SF_4, such as SF_4–HF–$Cl_2(Br_2)$ and SF_4–HF–S_2Cl_2, allows substitution of hydrogen by fluorine, and halofluorination of various classes of organic compounds.

In the present review, an attempt has been made to summarize the literature data referring chiefly to recent years, on the new uses of sulfur tetrafluoride in organic synthesis.

6.2   Reactions of Organic Compounds with Sulfur Tetrafluoride in Anhydrous Hydrogen Fluoride

Sulfur tetrafluoride is widely used to introduce fluorine into organic compounds. However substitution of carbonyl oxygen by fluorine using sulfur tetrafluoride in many cases fails to yields good results. A very important method of transforming

the aromatic carboxylic acids to trifluoromethyl-containing compounds possesses essential limitations. Thus, for example, the yields of benzotrifluoride and its
derivatives from benzoic acid and its analogues having electron-donating
substituents (CH_3, OCH_3) do not exceed 8–15% [25]. Even under severe
conditions (at 300 °C and above), polycarboxylic acids fail to give vicinal
poly(trifluoromethyl)benzenes [26]. The yields of fluorinated products are also
low in the reactions of SF_4 with carboxylic acids and ketones containing several
strong electron-accepting groups in the molecule.

An important achievement in sulfur tetrafluoride chemistry, substantially
expanding its preparative utility, was the modification of its reactions by adding
significant amounts (10 to 30 mol per mol of the carbonyl compound) of
anhydrous HF. This raised the yields of fluorinated products, and in many cases
allowed one to alter the reaction route.

6.2.1   Aromatic Carboxylic Acids

The use of excess hydrogen fluoride in the reactions of aromatic carboxylic acids
with SF_4 raises the yields of trifluoromethyl derivatives steeply both with
electron-donating and electron-accepting substituents, and enables the reactions
to take place under much milder conditions (Table 1).

Even in very severe conditions (300 °C and more), the reactions of SF_4 with
sterically hindered aromatic and heterocyclic polycarboxylic acids give only
poly(trifluoromethyl)benzoyl fluorides [26,31], while in HF they afford the
respective poly(trifluoromethyl)benzenes and poly(trifluoromethyl)furanes
[32,36] (Table 2).

The sterically hindered benzene-, naphthalene-, and furane polycarboxylic
acids with two carboxyl groups screened on both sides are transformed by SF_4 to
cyclic compounds containing the CF_2–O–CF_2 fragment. The reaction proceeds
via anhydride formation.

$X=Y=Br, NO_2$; $X=COOH$; $Y=CF_3$

Substitution of carbonyl oxygen by fluorine in anhydrides proceeds with
difficulty. The use of HF in these reactions allows one to obtain fluorinated
products in high yields (Table 3).

6.2.2   Esters of Carboxylic Acids

6.2.2.1   Reactions Forming Trifluoromethyl Derivatives

The reactions of esters of carboxylic acids with SF_4 give trifluoromethyl
derivatives and alkyl fluorides. These reactions require much more severe

Table 3. Reactions of benzene-, naphthalene-, and furanepolycarboxylic acids containing two sterically hindered carboxylic groups with SF_4 in HF

Acid	Ratio of acid : SF_4 : HF (mol)	T(°C)	Time (h)	Products	Yield (%)	Ref.
1,2,3,4-Benzenetetracarboxylic	1 : 20 : 0	200	14		76	31
3,6-Bis(trifluoromethyl)phthalic	1 : 10 : 0	20	12	3,6-**Bis**(trifluoromethyl)phthalic anhydride	92	37
3,6-Bis(trifluoromethyl)phthalic	1 : 10 : 0	200	18		74	37
3,6-Dibromobenzenetetracarboxylic	1 : 10 : 0	20	12	3,6-**Dibromo**-1,2:4,5-benzene-tetracarboxylic anhydride	92	37
3,6-Dibromobenzenetetracarboxylic	1 : 10 : 0	240–250	30		84	37

Starting acid	Ratio			Product			
3,6-Dinitrobenzenetetracarboxylic	1:20:50	160	15		41	33	
1,8-Naphthalenedicarboxylic	1:10:0	20	12	1,8-Naphthalenedicarboxylic anhydride	97	37	
1,8-Naphthalenedicarboxylic	1:10:0	220	10		63	37	
4,5-Dinitronaphthalene-1,8-di-carboxylic	1:6:0	260	12	4,5-Dinitronaphthalene-1,8-dicarboxylic anhydride	100	29	
4,5-Dinitronaphthalene-1,8-di-carboxylic	1:6:20	180	10		80	29	
3,6-Dinitronaphthalene-1,8-di-carboxylic	1:6:0	240	10		35	29	

Table 4. Pentafluoroethoxybenzenes $XC_6H_4OC_2F_5$ formed in the reaction of $XC_6H_4OCOCF_3$ with SF_4 [43]

X	Method	Yield (%)	X	Method	Yield (%)
H	A	61.0	4-CH_3	A	0
2-Cl	A	35.6	4-CH_3	B	45.6
2-Cl	B	71.4	2-CH_3O	A	0
3-Cl	A	45.7	2-CH_3O	B	35.6
4-Cl	A	46.3	2-NO_2	A	29.2
4-Cl	B	73.8	3-NO_2	A	69.0
4-F	A	36.7	4-NO_2	A	82.0
4-F	B	68.6	2,4-$(NO_2)_2$	A	19.3
2-CH_3	A	0	2,4-$(NO_2)_2$	B	55.8
2-CH_3	B	39.5	2-NO_2-4-Me	A	25.4
3-CH_3	A	0			
3-CH_3	B	62.5			

Method A. The reagents (ester: SF_4: HF = 1:2:1, mol) are heated to 100 to 175 °C for 6 h.
Method B. The reagents (ester: SF_4: HF = 1:2:10, mol) are kept at 20 °C for 15 h.

Using an excess of HF in the reactions of aryl trifluoroacetates with SF_4 allows easy synthesis of vicinal poly(pentafluoroethoxy)benzenes from sterically hindered trifluoroacetates of polyatomic phenols. Thus 1,2,3-tris(trifluoro-acetoxy)benzene heated with SF_4 at 150 °C in the presence of small additions of HF (mol per mol of ester) forms chiefly 2,6-bis(pentafluoroethoxy)phenyl trifluoroacetate, the yield of 1,2,3-tris(pentafluoroethoxy)benzene being only 11% [44]. In an excess of HF (10 mol per mol of ester), the latter is formed with a 60% yield already at 100 °C [45]. In the same manner, SF_4 reacts with 1,2,3,4-tetrakis(trifluoroacetoxy)benzene [45].

X=Y=H (60%); X=OCOCF$_3$, Y=OC$_2$F$_5$ (79%)

Likewise, 2,3,6-tris(pentafluoroethoxy)-4-nitrophenyl trifluoroacetate gives 2,3,5,6-tetrakis(pentafluoroethoxy)-4-nitrobenzene with a 66% yield, and 3,4,5,6-tetrakis(pentafluoroethoxy)phenylene 1,2-bis(trifluoroacetate) forms hexakis(pentafluoroethoxy)benzene.

Sulfur tetrafluoride in HF is widely used to prepare various perfluoroalkyl ethers, especially those containing strong electron-accepting substituents [47].

$$(CF_3)_3COCOC(CF_3)_3 + SF_4 \xrightarrow[250\ °C]{HF} (CF_3)_3COCF_2OC(CF_3)_3$$

Patent [48] describes difluoroformals containing nitro groups instead of perfluoroalkyl ones.

$$(CF(NO_2)_2CH_2O)_2C=O + SF_4 \xrightarrow{HF} (CF(NO_2)_2CH_2O)_2CF_2$$

In the reaction of SF_4 with esters of fluorine-containing glycols and perfluorodicarboxylic acids, the carbonyl oxygen of ester groups is substituted by fluorine atoms, and free carbonyl groups are only transformed to fluoro-formyl ones [63].

$$(HOCO(CF_2)_3COO)_2R_F + SF_4 \xrightarrow[150\,°C]{HF} (FCO(CF_2)_3CF_2O)_2R_F$$

The reactions of poly- and perfluoroacyloxy- derivatives of methyl benzoate with SF_4 in HF at 75 to 80 °C also produce polyfluoro- or perfluoroalkyl ethers, with the methoxycarbonyl group transformed to the trifluoromethyl one [49].

$$R_FCOOC_6H_4COOMe + SF_4 \xrightarrow{HF} R_FCF_2OC_6H_4CF_3$$

$$R_F = C_3F_7, H(CF_2)_4, C_4F_9, H(CF_2)_6, C_6F_{13}$$

6.2.3 Ketones

With ketones, as with carboxylic acids, the use of large amounts of HF in the reactions of SF_4 yields good results.

In the absence of HF, acetone reacts with SF_4 at 110 °C to form 2,2-difluoropropane with a 60% yield [2]. When this reaction is carried out in HF at 20 °C (6 to 8 mol of HF per mol of ketone), 2,2-difluoropropane is obtained in a quantitative yield. In a similar way SF_4 reacts with ethyl methyl ketone and diethyl ketone [50].

$$RCOR' + SF_4 \xrightarrow[20\,°C,\ 10\,h]{HF} RCF_2R'$$
$$98–100\%$$

$$R=R'=Me;\ R=R'=Et;\ R=Me,\ R'=Et$$

Accumulation of electron-accepting substituents in the ketone molecule deactivates the carbonyl group in the reaction with SF_4, and may be transformed to the difluoromethylene one only in rigid conditions in HF.

In this way perfluoroanthraquinone was converted to perfluoro-9,10-di-hydroanthracene [51].

groups. This is especially dramatically manifested in the aromatic series (Table 5).

The reactions of SF_4 with aliphatic dicarboxylic acids containing 2 or 3 carbons between the carboxylic groups, have their own distinctions (Table 6). Along with hexafluoroalkanes, the reaction gives no trifluoroalkanoyl fluorides as in the case of adipic and sebacic acids, but the isomeric or cyclic derivatives of 2,2,5,5-tetrafluorotetrahydrofurane and 2,2,6,6-tetrafluorotetrahydropyrane in a 30–70% yield. Besides, small amounts of linear $\alpha,\alpha,\alpha',\alpha'$-tetrafluoroalkyl ethers are formed in this reaction (Table 6) [61].

Cyclic tetrafluoroethers $(\underset{\underset{O}{\rule{1.2cm}{0.5pt}}}{CF_2(A)CF_2}$ may be used for the synthesis of trifluoroalkanoic acids, since upon heating with anhydrous HF they are transformed to trifluoroalkanoyl fluoride isomers [61].

$$HOOC(A)COOH + SF_4 \rightarrow CF_3(A)CF_3 + \underset{\underset{O}{\rule{1.2cm}{0.5pt}}}{CF_2(A)CF_2} + [CF_3(A)CF_2]_2O$$

$$\underset{\underset{O}{\rule{1cm}{0.5pt}}}{CF_2(A)CF_2} \underset{-H^+}{\overset{H^+}{\rightleftharpoons}} [^+CF_2(A)CF_2OH] \underset{-HF}{\overset{F^-}{\longrightarrow}} CF_3(A)COF$$

Thus 2,2,5,5-tetrafluorotetrahydrofurane is converted to 3,3,3-trifluorobutyryl fluoride, and 2,2,6,6-tetrafluorotetrahydropyrane—to 4,4,4-trifluoropentanoyl fluoride [61].

$$\underset{\underset{O}{\rule{1.5cm}{0.5pt}}}{CF_2(CH_2)_nCF_2} + HF \xrightarrow[100 \text{ to } 170\,°C]{} CF_3(CH_2)_nCOF$$

$$n = 2(52\%),\ 3(96\%)$$

The reactions of polyketones with SF_4 proceed stepwise. Therefore even in an excess of SF_4, 1,3-dibenzoylperfluoropropane may be converted to the octafluoro- and decafluoro-derivative [2].

$$PhCO(CF_2)_3COPh + 3SF_4 \xrightarrow[150 \text{ to } 220\,°C]{} \begin{cases} \xrightarrow[20\,h]{} PhCO(CF_2)_4Ph \\ \\ \xrightarrow[51\,h]{} Ph(CF_2)_5Ph \end{cases}$$

The reaction of SF_4 with acenaphthenequinone gives, depending on the amount of SF_4 used, 1,1-difluoroacenaphthenone or 1,1,2,2-tetrafluoroacenaphthene as the major reaction products [62].

Table 5. Reactions of aromatic dicarboxylic acids with SF_4

Acid HOOC(A)COOH	Ratio of SF_4:acid (mol)	T(°C)	Time (h)	Yield (%)			Ref.
				$CF_3(A)CF_3$	$CF_3(A)COF$	FOC(A)COF	
1,2-Benzenedicarboxylic	3:1	20	6	—	—	98	37
1,2-Benzenedicarboxylic	5:1	120	6	43	23	—	2
1,4-Benzenedicarboxylic	3:1	150	8	—	74	—	25
1,4-Benzenedicarboxylic	5:1	120	6	76	3	—	2
1-Trifluoromethylbenzene-2,6-dicarboxylic	6:1	250	30	17	60	—	32
Naphthalene-1,2-dicarboxylic	2:1	100	6	—	—	95	30
Naphthalene-1,2-dicarboxylic	3:1	100	6	—	90	—	30
Naphthalene-2,3-dicarboxylic	2:1	100	6	—	—	93	30
Naphthalene-2,3-dicarboxylic	3:1	100	6	—	92	—	30
Naphthalene-2,3-dicarboxylic	5:1	200	12	42	—	—	30
Naphthalene-2,6-dicarboxylic	2:1	100	6	—	—	98	30
Naphthalene-2,6-dicarboxylic	3:1	100	10	—	95	—	30
Furan-3,4-dicarboxylic	4:1[a]	100–115	12	70	—	—	60
Furan-3,4-dicarboxylic	2.8:1[b]	100	10	—	79	—	60

[a] With 2 moles of HF
[b] With 10 moles of HF

Table 7. Reactions of sterically hindered benzenepolycarboxylic acids with SF_4

Acid	Ratio of SF_4:acid (mol)	$T(°C)$	Time (h)	Product	Yield (%)	Ref.
3-Nitrobenzene-1,2-dicarboxylic	4.5:1	100–140	8	3-Nitro-6-trifluoromethyl-benzoyl fluoride	85	26
1,2,3-Benzenetricarboxylic	8:1	150–175	15	2,6-Bis(trifluoromethyl)-benzoyl fluoride	84	26
1,2,3-Benzenetricarboxylic	4:1	120–150	11	3-Trifluoromethylphthaloyl difluoride	69	63
1,2,3,4-Benzenetetracarboxylic	6:1	120–170	10	3,6-Bis(trifluoromethyl)-phthaloyl difluoride	76	63
1,2,3,5-Benzenetetracarboxylic	10:1	150–200	14	2,4,6-Tris(trifluoromethyl)-benzoyl fluoride	79	31
3-Trifluoromethyl-1,2,4-benzenetricarboxylic	10:1	150–220	19	2,3,6-Tris(trifluoromethyl)-benzoyl fluoride	76	37
2,5,6-Tris(trifluoromethyl)-1,3-benzenedicarboxylic	8:1	290	15	2,3,5,6-Tetrakis(trifluoromethyl)benzoyl fluoride	82	64

Benzenepolycarboxylic acids having at least four vicinal carboxyl groups possess a strong steric hindrance for the transformation of middle COOH group to the trifluoromethyl ones. Under the mild conditions as seen from Table 7, this leads to 3,6-bis(trifluoromethyl)phthaloyl difluoride in a high yield. Upon heating with SF_4 in more stringent conditions, the trifluoromethyl derivatives of 1,1,3,3-tetrafluorophthalane are formed [31,65].

X = Y = H (76 %), X = COOH, Y = CF$_3$ (54 %)

Benzenehexacarboxylic acid reacts with SF_4, also giving a mixture of cyclic compounds containing the CF_2OCF_2 group [37,65].

The difluoromethylene groups of the tetrafluorophthalane cycle are extremely stable against alkaline solutions, but are more easily hydrolysed in acid solutions than the trifluoromethyl ones. Hence treatment of trifluoromethyl-containing tetrafluorophthalanes with concentrated H_2SO_4 affords the respective poly(trifluoromethyl)phthalic acids in high yields [37,65].

The reaction of naphthalene-1,8-dicarboxylic acid and its derivatives with SF_4 give no bis(trifluoromethyl)naphthalenes but the derivatives of 1,1,3,3-tetrafluoro-1H-naphtho[1,8-c,d]-pyrane [29,37].

Table 8. Reactions of hydroxybenzoic acids with SF_4 [66]

Acid	Ratio of acid : SF_4 : HF (mol)	T(°C)	Time (h)	Product	Yield (%)
5-Bromo-2-hydroxybenzoic	1:4:25	25	15	5-Bromo-2-hydroxybenzotri-fluoride	45
5-Nitro-2-hydroxybenzoic	1:5:24	25	12	5-Nitro-2-hydroxybenzotri-fluoride	73
3-Hydroxybenzoic	1:4:25	20	15	3-Hydroxybenzotrifluoride	75
4-Hydroxybenzoic	1:4:25	10	90	4-Hydroxybenzotrifluoride	90
3,5-Dinitro-4-hydroxybenzoic	1:5:25	10	80	3,5-Dinitro-4-hydroxybenzo-trifluoride	92

In the same manner SF_4 reacts with 1-methyl-5-formyluracyl [72]. The 4,6-dihydroxy-1,2,5-triazine-3-carboxylic acid is transformed by SF_4 to 4,6-dihydroxy-3-trifluoromethyl-1,2,5-triazine [73].

The selective fluorination of polyketones with SF_4 was demonstrated by Shirota and his co-workers [74]. The oxo groups at 2- and 5-positions of tetrone are easily converted to the difluoromethylene ones, the other keto groups remain intact [74].

5-Carboxy-2-pyrone also reacts with SF_4 with preservation of the oxo group [69].

The oxygen-containing functional groups conjugated with the unsaturated bonds seem to be less active than the non-conjugated ones. This is vividly seen in the reactions of unsaturated dicarboxylic acids with SF_4 affording the products of substitution of the trifluoromethyl group only for one of the two carboxyl groups. Thus the reaction of SF_4 with itaconic and 1-methyl-1-cyclobutene-2,3-dicarboxylic acids may give the products in which the carboxyl group adjacent to the carbon atom of the double bond forms only the fluoroformyl bond [2].

The same reasons account for the low reactivity of the conjugated 3-oxo group in α,β-unsaturated steroids, whereas in the saturated analogues it is substituted easier by the difluoromethylene group in the reaction with SF_4 than other groups.

The reactivity of the keto groups at different carbon atoms of the steroid frame decreases in the series: 3-keto > 6-keto > 17-keto > 20-keto-11-keto-steroid > 20-keto-11-deoxysteroid > conjugated 3-keto > 11-keto [56,57]. The aldehyde group of steroids is very easily transformed to the difluoromethyl one. Table 9 lists the results of the selective reactions of di- and triketosteroids and aldosteroids.

$$CF_3CH-CH_2 + SF_4 \xrightarrow{NaF} CF_3CH(OSOF)CH_2F \xrightarrow{H_2O} CF_3CHOHCH_2F$$
$$\underset{OH \ OH}{\quad} \qquad\qquad 98\%$$

The reactions of SF_4 with compounds having vicinal hydroxy groups may be exemplified by the reactions of SF_4 with esters of tartaric acids [90–92]. Heating of dimethyl (+)- and (−)-tartrates with SF_4 leads to a mixture of *erythro*-2-fluoro-1,2-bis(methoxycarbonyl)ethyl fluorosulfite *1*, dimethyl (−)(2S:3S)-2-fluoro-3-hydroxysuccinate *2*, and dimethyl *meso*-2,3-difluoro-succinate *3*, the components being the same upon variation of the reaction temperature from 50 to 180 °C. The reaction proceeds with complete reversal of configuration at one of carbon atoms, and with its complete preservation at the second carbon atom [91,92].

The authors of [92] found that the hydroxyester *2* is formed upon treatment of fluorosulfite *1* with HF evolving in the reaction of dimethyl tartrate with SF_4. The subsequent substitution of the hydroxy group by fluorine under the action of SF_4 in the presence of HF leads to the difluoroester *3*. To obtain mono-fluoromaloate *2*, the reaction with SF_4 is carried out at 20 °C in the presence of an HF acceptor—sodium fluoride. Fluorosulfite *1* is obtained here in a quantitative yield and after hydrolysis gives the hydroxyester *2*. The product of substitution of fluorine for both hydroxy groups—dimethyl difluorosuccinate *3* may also be obtained in a quantitative yield in the reaction of dimethyl tartrate with SF_4 in excess HF.

The intermediate formation of difluorosulfuranes and fluorosulfites seems to be also characteristic for the reactions of 1,3- and 1,4-diols. However these compounds were not isolated. The reactions of 1,3-propanediol and 1,3-butanediol with SF_4 gave the respective fluorine-containing alcohols.

$$HOCH_2CH_2CH_2OH + SF_4 \rightarrow CH_2FCH_2CH_2OH$$

$$CH_3CHOHCH_2CH_2OH + SF_4 \rightarrow CH_3CHFCH_2CH_2OH$$

Treatment of 1,4-butanediol with SF_4 yields, along with 4-fluoro-1-butanol, an appreciable amount of tetrahydrofuran formed by dehydration.

$$HOCH_2(CH_2)_2CH_2OH \; + \; SF_4 \longrightarrow CH_2F(CH_2)_2CH_2OH \; + \; \begin{matrix} CH_2-CH_2 \\ | \qquad | \\ CH_2 \quad CH_2 \\ \diagdown O \diagup \end{matrix}$$

Accumulation of electron-accepting groups in the glycol molecule increases the stability of carbon–oxygen bonds to such an extent that these bonds do not undergo cleavage under the action of the fluoride ion. Thus perfluoropinacone reacts with SF_4 to give a quantitative yield of perfluorinated spirosulfurane 4 [93].

$$\begin{matrix} (CF_3)_2C-OH \\ | \\ (CF_3)_2C-OH \end{matrix} \; + \; SF_4 \; \xrightarrow{20\,°C} \; \begin{matrix} (CF_3)_2C-O \diagdown \quad \diagup O-C(CF_3)_2 \\ | \qquad\qquad S \qquad | \\ (CF_3)_2C-O \diagup \quad \diagdown O-C(CF_3)_2 \end{matrix}$$

$$4 \qquad 100\,\%$$

Accumulation of the hydroxy groups in the molecules of polyatomic alcohols leads to a sharp decrease of the yield of fluorine-containing compounds due to the strong resinification and carbonisation of the products. This may be avoided by carrying out the reactions in anhydrous HF. In this case the reaction of glycerol with SF_4 shows the regularities found for the reactions of SF_4 with diatomic alcohols. Thus the reaction of SF_4 with glycerol in HF at low temperatures gave 3-fluoro-1,2-propylene sulfite in a high yields, which seems to be formed as a result of hydrolysis of the intermediate difluorosulfurane C [80].

$$\begin{matrix} CH_2OH \\ | \\ CHOH \\ | \\ CH_2OH \end{matrix} \; + \; SF_4 \; \xrightarrow[-40\,°C]{HF} \; \left[\begin{matrix} CH_2-O \diagdown \\ | \qquad\quad SF_2 \\ CH-O \diagup \\ | \\ CH_2OSF_3 \end{matrix} \right] \; \xrightarrow[\substack{-SOF_2^+ \\ -F^-}]{HF} \; \left[\begin{matrix} CH_2-O \diagdown \\ | \qquad\quad SF_2 \\ CH-O \diagup \\ | \\ CH_2F \end{matrix} \right] \longrightarrow \begin{matrix} CH_2-O \diagdown \\ | \qquad\quad S=O \\ CH-O \diagup \\ | \\ CH_2F \end{matrix}$$

$$C \qquad\qquad D \quad 82\,\%$$

In more rigid conditions, at 20 °C, difluorosulfurane C undergoes the F^- attack to form fluorosulfite E which is transformed in HF to 1,3-difluoro-2-propanol.

$$C + HF \longrightarrow CH_2FCH(OSOF)CH_2F \xrightarrow{HF} CH_2FCHOHCH_2F$$

$$E$$

Similar results have been obtained for D-serine. On the basis of these data, the authors conclude that in the threonine–*allo*-threonine system the reactions follow the S_N2 mechanism.

On the other hand, fluorodeoxylation of 2-methylserine gives, along with the expected 2-(fluoromethyl)alanine, 40% of 1-amino-cyclopropanecarboxylic acid.

$$\underset{\underset{NH_2}{|}}{\overset{\overset{CH_3}{|}}{HOCH_2CCOOH}} + SF_4 \xrightarrow{\ HF\ } \underset{\underset{NH_2}{|}}{\overset{\overset{CH_3}{|}}{FH_2CCCOOH}} + \begin{array}{c} H_2C \diagdown \quad \diagup NH_2 \\ | \quad C \\ H_2C \diagup \quad \diagdown COOH \end{array}$$

Formation of the latter is explained by the fact that the reactions proceed by the S_N1 mechanism via the carbonium ion. The reactive primary carbonium ion is incorporated at the C–H bond of the CH_3 group.

The authors did not manage to carry out deoxylation of diastereomeric 3-hydroxyaspartic acids, whereas the reactions of their esters readily proceed at $0\,^\circ C$.

$$MeOCOCHOHCH(NH_2)COOMe + SF_4 \xrightarrow[0\,^\circ C]{HF}$$

$$MeOCOCHFCH(NH_2)COOMe$$

Fluorodeoxylation of hydroxyamino acids by SF_4 may be improved by carrying out the reaction in HF and in the presence of BF_3 or $AlCl_3$ which markedly facilitate substitution of the OH group by fluorine. This method affords high yields of α-fluoromethylamino acids from the respective hydroxy-derivatives [91,92].

$$\underset{\underset{NH_2}{|}}{\overset{\overset{CH_2OH}{|}}{RCH_2CCOOH}} + SF_4 \xrightarrow{HF,\ BF_3\ or\ AlCl_3} \underset{\underset{NH_2}{|}}{\overset{\overset{CH_2F}{|}}{RCH_2CCOOH}}$$

$$R = C_6H_5,\ HOC_6H_4,\ (HO)_2C_6H_3,\ NH_2CH_2CH_2$$

6.3.4.6   Aliphatic Dialdehydes and Diketones

Glyoxal [93], diacetyl [94], and perfluorodiacetyl [95] easily react with SF_4 in mild conditions to form fluorine-containing spiro-sulfuranes.

$$\underset{RC-CR}{\overset{O\ \ O}{\overset{||\ ||}{}}} + SF_4 \longrightarrow \begin{array}{c} RFC-O \diagdown \quad \diagup O-CFR \\ | \qquad S \qquad | \\ RFC-O \diagup \quad \diagdown O-CFR \end{array}$$

R=H (50 %), CH_3 (35%), CF_3 (100%)

Asymmetric 1,2-diketones react with SF_4 to give mixtures of tetrafluoro-alkanes and fluoroolefins [94].

$$RCOCOCH_2R' + SF_4 \rightarrow RCF_2CF_2CH_2R' + RCF_2CF{=}CHR'$$
$$\qquad\qquad\qquad\quad\ 60\text{–}65\% \qquad\qquad 25\text{–}30\%$$

6.3.4.7   Aliphatic Oxo Carboxylic Acids and Their Esters

Treatment of 2-oxo carboxylic acids with SF_4 leads to decarboxylation and formation of trifluoroalkanes containing one carbon less than the molecule of the starting acid [96].

$$RCOCOOH + SF_4 \rightarrow RCF_3$$

$$R = Me, \ Ph$$

The reaction of 2-oxoglutaric acid with SF_4 [96] gives the same products as the reaction of SF_4 with succinic acid [61], but the yield of 2,2,5,5-tetrafluoro-tetrahydrofuran is twice as high.

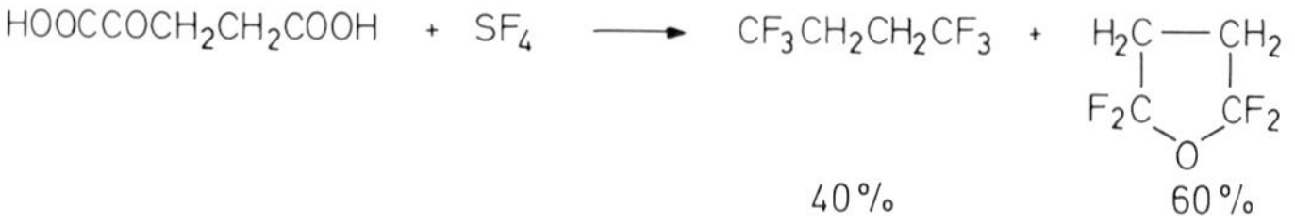

$$HOOCCOCH_2CH_2COOH \ + \ SF_4 \longrightarrow CF_3CH_2CH_2CF_3 \ + \ \underset{F_2C \diagdown_O\diagup CF_2}{\overset{H_2C - CH_2}{|\qquad|}}$$

$$40\% \qquad\qquad 60\%$$

The selective substitution of the oxo group by two fluorine atoms with preservation of the carboxyl group is only feasible in some 3-oxo dicarboxylic and 4-oxo carboxylic acids. Thus 3-oxo-glutaric acid reacts with SF_4 at 100 °C to give 3,3,5,5-tetrafluoropentanoyl fluoride, or in HF–1,1,1,3,3,5,5,5-octafluoro-pentane [96].

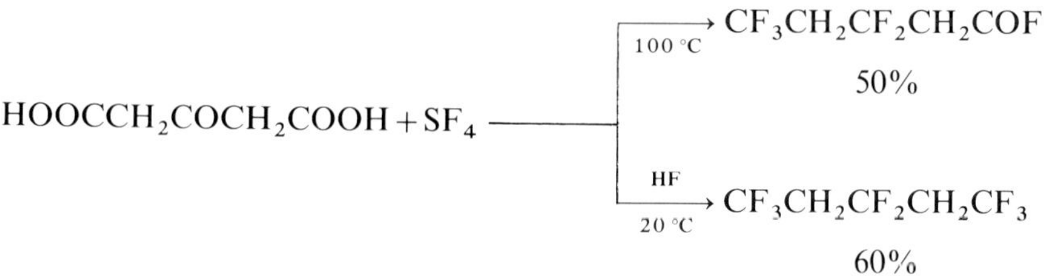

$$HOOCCH_2COCH_2COOH + SF_4 \quad \begin{cases} \xrightarrow[100\ °C]{} CF_3CH_2CF_2CH_2COF \\ \qquad\qquad\qquad 50\% \\[2em] \xrightarrow[20\ °C]{HF} CF_3CH_2CF_2CH_2CF_3 \\ \qquad\qquad\qquad 60\% \end{cases}$$

The varying reactivity of carbonyl groups seems to be due to the presence of enol forms in the starting acid. Enolization accounts for the fact that , in contrast to 2-oxoglutaric acid—which as shown above releases CO in the reaction with SF_4— the oxomalonic acid gives, along with decarbonylation products, compounds where the oxo group is substituted by two fluorines [96].

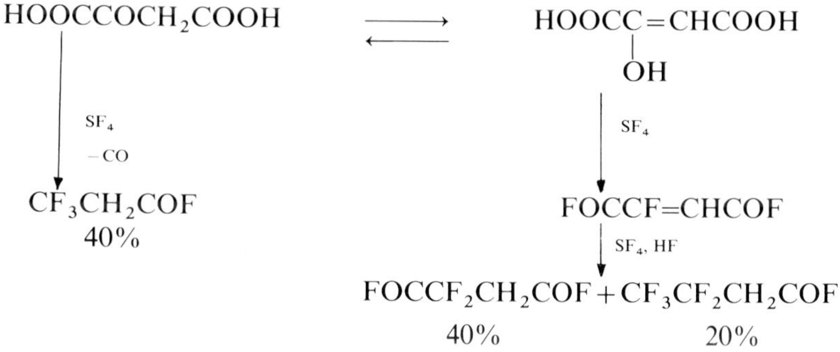

$$\begin{array}{ccc} HOOCCOCH_2COOH & \xrightleftharpoons{\qquad} & \underset{\quad\ \ OH}{HOOCC=CHCOOH} \\ \Big\downarrow {\scriptstyle SF_4 \atop -CO} & & \Big\downarrow {\scriptstyle SF_4} \\ CF_3CH_2COF & & FOCCF=CHCOF \\ 40\% & & \Big\downarrow {\scriptstyle SF_4,\ HF} \\ & FOCCF_2CH_2COF + CF_3CF_2CH_2COF \\ & 40\% \qquad\qquad\qquad 20\% \end{array}$$

S. The latter reacts with the fluoride ion, giving fluoroalkyldisulfenyl chloride 29 which reacts in a similar way with the next propene molecule to form the disulfide 28 [50].

$$CH_3CH=CH_2 + SF_4 + S_2Cl_2$$

$$\xrightarrow{\text{HF}} [CH_3\overset{+}{C}HCH_2SSCl \xrightarrow{\text{F}^-} CH_3CHFCH_2SSCl]$$
$$ S 29$$

$$29 + SF_4 + S_2Cl_2 + CH_3CH=CH_2 \rightarrow CH_3CHFCH_2SSCH_2CHFCH_3$$
$$ 28$$

The addition of stoichiometric equivalents of "S_2F_2" to propene proceeds according to the Markownikov rule. Substitution of the methyl group of propene by the trifluoromethyl one completely changes orientation of the addition. 3,3,3-Trifluoropropene reacts with SF_4–HF–S_2Cl_2, forming a mixture of two stereo-isomeric sulfides 30 [132].

$$CF_3CH=CH_2 + SF_4 + S_2Cl_2 \xrightarrow{\text{HF}} (CH_2FCH)_2S$$
$$ \underset{F_3C}{|}$$
$$ 30$$

Formation of sulfides but not disulfides in this reaction, as in the reaction of propene, is explained in the following way. 3,3,3-Trifluoropropene is initially transformed to thiosulfene chloride 31 which undergoes cleavage by chlorine to give sulfene chloride 32. The reaction of the latter with the next 3,3,3-trifluoropropene molecule leads to sulfides 30 [132].

$$CF_3CH=CH_2 + SF_4 + S_2Cl_2 \xrightarrow{\text{HF}} [^+CH_2CH(CF_3)SSCl$$

$$\xrightarrow{\text{F}^-} CH_2FCH(CF_3)SSCl \xrightarrow[-SCl_2]{Cl_2} CH_2FCH(CF_3)SCl]$$
$$ 32 31$$

$$\xrightarrow{CF_3CH=CH_2,\ SF_4,\ HF} 30$$

Perhaloolefins react with SF_4–HF–S_2Sl_2 under more severe conditions than 3,3,3-trifluoropropene. In this case the reaction stops at the stage of formation of sulfene chlorides. 1,2-Dichlorodifluoroethylene reacts with SF_4–HF–S_2Cl_2 at temperatures above 60 °C, giving 1,2-dichloroperfluoroethylsulfur chloride.

$$CFCl=CFCl + SF_4 + S_2Cl_2 \xrightarrow[60\,°C]{\text{HF}} CF_2ClCFClSSCl$$

$$\xrightarrow[-SCl_2]{Cl_2} CF_2ClCFClSCl$$
$$ 74\%$$

Chlorotrifluoroethylene reacts with SF_4–HF–S_2Cl_2 to form a mixture of regioisomeric chlorotetrafluoroethylsulfur monochlorides *33* and *34*.

$$CF_2-CFCl + SF_4 + S_2Cl_2 \xrightarrow{\text{HF}} CF_3CFClSCl + CF_2ClCF_2SCl$$

$$\qquad\qquad\qquad\qquad\qquad\qquad\quad \textit{33}, 28\% \qquad\quad \textit{34}, 39\%$$

As shown above, aliphatic ketones and difluoroalkanes react with SF_4–HF–Cl_2, yielding vicinal chlorodifluoroalkanes. In the reaction of ketones and difluoroalkanes with SF_4–HF–S_2Cl_2, no chlorodifluoroalkanes are formed; depending on the temperature, the reaction gives polyfluoroalkyl sulfides and polyfluoroalkyl disulfides. The reaction products always contain molecular sulfur.

Acetone and 2,2-difluoropropane are transformed by SF_4–HF–S_2Cl_2, depending on conditions, to polyfluoroalkyl sulfides *35*, *36*, or disulfide *37* [50,133].

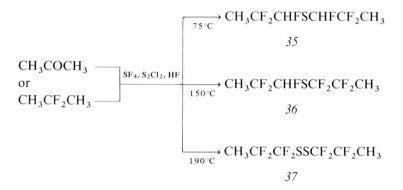

2,2-Difluoropropane formed at the first stage of the reaction of acetone with SF_4–HF reacts with the F_3S^+ ion with F^- elimination to form the carbocation T which releases the proton to give 2-fluoropropene. The latter reacts with SF_4–HF–S_2Cl_2 to yield 2,2,2′,2′-tetrafluorodipropyl sulfide. Its chlorination and subsequent substitution of chlorine by fluorine leads to sulfides *35*.

$$CH_3COCH_3 + SF_4 \xrightarrow{\text{HF}} CH_3CF_2CH_3$$

$$\xrightarrow{+SF_3} [CH_3\overset{+}{C}FCH_3 \xrightarrow{-H^+} CH_3CF=CH_2$$
$$\qquad\qquad T$$

$$\xrightarrow{SF_4,\ S_2Cl_2,\ HF} (CH_3CF_2CH_2)_2S] \rightarrow 35$$

Ethyl methyl ketone is converted by SF_4–HF–S_2Cl_2 to the sulfide *38* and the disulfide *39*. The same products are formed from 2,2-difluorobutane. The yield of

5. Kondratenko NV, Syrova GP, Sheinker YuN, Popov VI, Yagupolskii LM (1971) Zh. Obshch. Khim. 41:2056
6. Yagupolskii LM, Kondratenko NV, Popov VI (1976) Zh. Obshch. Khim. 46:620
7. Yagupolskii LM, Lyalin VV, Orda VV (1968) Zh. Obshch. Khim. 38:2813
8. Lyalin VV, Syrova GP, Orda VV, Alekseeva LA, Yagupolskii LM (1970) Zh. Org. Khim. 6:1420
9. Harder RJ, Smith WC (1961) J. Amer. Chem. Soc. 83:3422
10. US Pat 2862029 (1959); (1959) Chem. Abs. 53:9152
11. Smith WC, Tallock CW (1960) J. Amer. Chem. Soc. 52:551
12. Yagupolskii LM, Burmakov AI, Alekseeva LA (1971) in: Reakzii i metody issledovaniya organicheskikh soedinenii, vol 22. Khimiya, Moskva, p 40
13. Boswell GA, Ripka WC (1973) Org. Reactions 21:1
14. Khardin AP, Gorbunov BN, Protopopov PA (1973) Khimiya chetyrekhftoristoi sery. Saratovskii universitet, Saratov, p 201
15. Martin DC (1967) An. New York Acad. Sci. 145:161
16. US Pat 2992073 (1961); (1961) Chem. Abs. 55:27813
17. Tallock CW, Fawcett FS, Smith WC, Gofman DD (1960) J. Amer. Chem. Soc. 82:539
18. Fawcett FS, Tallock CW (1963) Inorg. Synth. 7:119
19. Franz R (1980) J. Fluor. Chem. 15:423
20. US Pat 3399063 (1958); (1958) Chem. Abs. 53:78864
21. Naumann D, Padma DK (1973) Z. anorg. allg. Chem. 401:53
22. Becher W, Massone J (1974) Chem. Ztg. 98:117
23. Markovskii LN, Pashinnik VE, Kirsanov AV: Synthesis 1973:787
24. Middleton JW (1975) J. Org. Chem. 40:574
25. Burmakov AI, Alekseeva LA, Yagupolskii LM (1973) Zh. Org. Khim. 8:153
26. Yagupolskii LM, Burmakov AI, Alekseeva LA (1969) Zh. Obshch. Khim. 39:2053
27. Kunshenko BV, Burmakov AI, Alekseeva LA, Lukmanov VG, Yagupolskii LM (1974) Zh. Org. Khim. 10:886
28. Kunshenko BV, Alekseeva LA, Yagupolskii LM (1972) Zh. Org. Khim. 8:830
29. Kunshenko BV, Alekseeva LA, Yagupolskii LM (1974) Zh. Org. Khim. 10:1698
30. Kunshenko BV, Alekseeva LA, Yagupolskii LM (1973) Zh. Org. Khim. 9:1954
31. Burmakov AI, Alekseeva·LA, Yagupolskii LM (1970) Zh. Org. Khim. 6:144
32. Lukmanov VG, Alekseeva LA, Burmakov AI, Yagupolskii LM (1973) Zh. Org. Khim. 9:1019
33. Lukmanov VG, Alekseeva LA, Yagupolskii LM (1977) Zh. Org. Khim. 13:2129
34. Lukmanov VG, Alekseeva LA, Yagupolskii LM (1974) Zh. Org. Khim. 10:2000
35. Lyalin VV, Grigorash RV, Alekseeva LA, Yagupolskii LM (1975) Zh. Org. Khim. 11:460
36. Grigorash RV, Lyalin VV, Alekseeva LA, Yagupolskii LM (1978) 14:2623
37. Yagupolskii LM, Burmakov AI, Alekseeva LA, Kunshenko BV (1973) Zh. Org. Khim. 9:689
38. Fialkov YuA, Moklyachuk LI, Kremlev MM, Yagupolskii LM (1980) Zh. Org. Khim. 16:1476
39. US Pat 4237276 (1980)
40. Sheppard WA (1961) J. Amer. Chem. Soc. 83:4860
41. Sheppard WA (1964) J. Org. Chem. 29:1
42. Aldrich PE, Sheppard WA (1964) J. Org. Chem. 29:11
43. Belous VM, Alekseeva LA, Yagupolskii LM (1975) Zh. Org. Khim. 11:1672

44. Alekseeva LA, Belous VM, Yagupolskii LM (1974) Zh. Org. Khim. 10:1053
45. Yagupolskii LM, Belous VM, Alekseeva LA (1976) Zh. Org. Khim. 12:1287
46. Belous VM, Litvinova KD, Alekseeva LA, Yagupolskii LM (1976) Zh. Org. Khim. 12:1798
47. Pasquale (1973) J. Org. Chem. 38:3025
48. US Pat 4210710 (1979)
49. US Pat 4201876 (1981)
50. Muratov NN, Mokhamed NM, Kunshenko BV, Alekseeva LA, Yagupolskii LM (1985) Zh. Org. Khim. 21:1420
51. Bardon J, Childs AC, Parsons LM: J. Chem. Soc. Chem. Commun. 1982:534
52. Yagupolskii LM, Kunshenko BV, Alekseeva LA (1968) Zh. Obshch. Khim. 38:2592
53. Gopal N, Snyder CE, Tamborski C (1979) J. Fluor. Chem. 14:511
54. Conlin RT, Frey NM: J. Chem. Soc. Faraday Trans. 1979:2556
55. Ishikawa N, Kitazume T, Takaoka A (1979) J. Synth. Org. Chem. Jap. 37:606
56. Martin DG, Kagan F (1962) J. Org. Chem. 27:3164
57. Tadanier J, Cole W (1961) J. Org. Chem. 26:2436
58. Boswell GA (1966) J. Org. Chem. 31:991
59. Azeem M, Brownstein M (1969) Can. J. Chem. 47:4159
60. Lyalin VV, Grigorash RV, Alekseeva LA, Yagupolskii LM (1975) Zh. Org. Khim. 11:1086
61. Dmowski W, Kolinski A (1978) Polish J. Chem. 52:71
62. Kunshenko BV, Alekseeva LA, Yagupolskii LM (1970) Zh. Org. Khim. 6:1286
63. Burmakov AI, Alekseeva LA, Yagupolskii LM (1969) Zh. Org. Khim. 5:1892
64. Yagupolskii LM, Lukmanov VG, Boiko VN, Alekseeva LA (1977) Zh. Org. Khim. 13:2388
65. Burmakov AI, Alekseeva LA, Yagupolskii LM (1970) Zh. Org. Khim. 6:2498
66. Blakitnii AN, Zalesskaya IM, Kunshenko BV, Fialkov YuA, Yagupolskii LM (1977) Zh. Org. Khim. 13:2149
67. Alekseeva LA, Belous VM, Lozinskii MO, Shendrik VP, Yagupolskii LM (1983) Ukr. Khim. Zh. 49:74
68. US Pat 2894989 (1959); (1961) Chem. Abs. 55:420
69. US Pat 4249009 (1981)
70. Mertes MP, Saheb SE (1963) J. Pharm. Sci. 52:508
71. Mertes MP, Saheb SE (1963) J. Med. Chem. 6:619
72. Sakai TT, Santi DV (1973) J. Med. Chem. 16:1079
73. US Pat 3324126 (1967); (1968) Chem. Abs. 68:39646
74. Shirota FN, Nagasawa HT, Elberling JA (1977) J. Med. Chem. 20:1176
75. Burmakov AI, Khassanein SM, Kunshenko BV, Alekseeva LA, Yagupolskii LM (1986) Zh. Org. Khim. 22:1273
76. Kozlova AM, Sedova LN, Alekseeva LA, Yagupolskii LM (1973) Zh. Org. Khim. 9:1418
77. Bell HH, Hudlicky M (1980) J. Fluor. Chem. 15:191
78. Burmakov AI, Motnyak LA, Kunshenko BV, Alekseeva LA, Yagupolskii LM (1981) J. Fluor. Chem. 19:151
79. Kryukova LYu, Kryukov LN, Kolomietz AF, Sokolskii GA, Knunyants IL: Izv. Akad. Nauk SSSR. Ser. Khim. 1979:1913
80. Khasanein SM, Burmakov AI, Bloshchiza FA, Yagupolskii LM (1987) Zh. Org. Khim. 22:1111
81. Stepanov IV, Burmakov AI (1985) Zh. Org. Khim. 21:45

82. Stepanov IV, Burmakov AI, Alekseeva LA, Yagupolskii LM (1986) Zh. Org. Khim. 22:227
83. Burmakov AI, Motnyak LA, Kunshenko BV, Alekseeva LA, Yagupolskii LM (1980) Zh. Org. Khim. 16:1401
84. Motnyak LA, Burmakov AI, Kunshenko BV, Sass VP, Alekseeva LA, Yagupolskii LM (1981) Zh. Org. Khim. 17:728
85. Motnyak LA, Burmakov AI, Kunshenko BV, Neizvestnaya TA, Alekseeva LA, Yagupolskii LM (1983) 19:720
86. Muratov NN, Burmakov AI, Kunshenko BV, Alekseeva LA, Yagupolskii LM (1982) Zh. Org. Khim. 18:1403
87. Motnyak LA, Burmakov AI, Kunshenko BV, Neizvestnaya TA, Alekseeva LA, Yagupolskii LM (1984) Zh. Org. Khim. 20:1169
88. Kollonitsch J, Marburg S, Perkins L (1975) J. Org. Chem. 40:3808
89. Kollonitsch J (1978) Isr. J. Chem. 17:53
90. US Pat 4096180 (1978)
91. US Pat 4215221 (1980)
92. US Pat 4288601 (1981)
93. Stepanov IV, Burmakov AI, Kunshenko BV, Alekseeva LA, Yagupolskii LM (1982) Zh. Org. Khim. 22:1812
94. Burmakov AI, Stepanov IV, Kunshenko BV, Sedova LN, Alekseeva LA, Yagupolskii LM (1982) Zh. Org. Khim. 18:1168
95. Hodges KS, Schomburg D, Weise, Schmutzler R (1977) J. Amer. Chem. Soc. 99:6096
96. Bloshchiza FA, Burmakov AI, Kunshenko BV, Alekseeva LA, Yagupolskii LM (1985) Zh. Org. Khim. 21:1414
97. Bloshchiza FA, Burmakov AI, Kunshenko BV, Alekseeva LA, Bel'ferman AL, Pazderskii YuA, Yagupolskii LM (1981) Zh. Org. Khim. 17:1417
98. Bloshchiza FA, Burmakov AI, Kunshenko BV, Alekseeva LA, Yagupolskii LM (1982) Zh. Org. Khim. 18:782
99. Bloshchiza FA, Burmakov AI, Kunshenko BV, Alekseeva LA, Yagupolskii LM (1986) Zh. Org. Khim. 22:750
100. Bloshchiza FA, Burmakov AI, Kunshenko BV, Alekseeva LA, Yagupolskii LM (1983) Zh. Org. Khim. 19:1761
101. Kul'chzkii MM, Il'chenko AYa, Yagupolskii LM (1973) Zh. Org. Khim. 9:827
102. Lack RE, Ganter C, Roberts JD (1968) J. Amer. Chem. Soc. 90:7001
103. Stepanov IV, Burmakov AI, Kunshenko BV, Alekseeva LA, Yagupolskii LM (1983) Zh. Org. Khim. 19:273
104. Kaufmann H, Fuher H, Kavloda T (1981) Tetrahedron 37:225
105. Burmakov AI, Bloshchiza FA, Kunshenko BV, Alekseeva LA, Yagupolskii LM (1980) Zh. Org. Khim. 16:2617
106. Appleguist DE, Searle R (1964) J. Org. Chem. 29:987
107. US Pat 4431786 (1983)
108. Spasov SL, Griffith DL, Glazer ES, Nagarayan K, Roberts JD (1967) J. Amer. Chem. Soc. 89:88
109. Khardin AN, Popov AD, Protopopov PA (1977) Zh. Vses. Khim. O-va 22:116
110. Aleksandrov AM, Danilenko TI, Konovalov EV, Krasnoshchek AN, Medvedeva TP (1974) Zh. Org. Khim. 10:1548
111. Aleksandrov AM, Kukhar' VP, Danilenko GI, Krasnoshchek AP (1977) Zh. Org. Khim. 13:1629
112. Aleksandrov AM, Sorochinskii AE, Krasnoshchek AP (1979) Zh. Org. Khim. 15:336

113. Dmowski W, Kaminski M (1983) J. Fluor. Chem. 23:207
114. Kunshenko BV, Muratov NN, Burmakov AI, Alekseeva LA, Yagupolskii LM (1983) J. Fluor. Chem. 22:105
115. Muratov NN, Kunshenko BV, Burmakov AI, Alekseeva LA, Yagupolskii LM (1984) Zh. Org. Khim. 20:450
116. Padma DK (1977) Phosphorus and Sulphur 3:19
117. Knunyants IL, German LS: Izv. Akad. Nauk SSSR. Ser. Khim. 1966:1065
118. Pattison FLM, Peters DAV, Dean FH (1965) Can. J. Chem. 43:1689
119. Olah GA, Nojima M, Kerekes I: Synthesis 1973:780
120. Muratov NN, Omarov VO, Kunshenko BV, Burmakov AI, Alekseeva LA, Yagupolskii LM (1986) Zh. Org. Khim. 22:1806
121. Sargent PB, Krespan CG (1969) J. Amer. Chem. Soc. 91:415
122. Dear REA, Gilbert EE, Murray JJ (1971) Tetrahedron 27:3345
123. Kunshenko BV, Motnyak LA, Neizvestnaya TA, Il'nitzkii SO, Yagupolskii LM (1986) Zh. Org. Khim. 22:1791
124. Grigorash RV, Lyalin VV, Alekseeva LA, Yagupolskii LM (1978) Zh. Org. Khim. 14:844
125. Sherman WR, Freiffider M, Stone GR (1960) J. Org. Chem. 25:2048
126. Grigorash RV, Lyalin VV, Alekseeva LA, Yagupolskii LM: Khim. geterozikl. soed. 1977:1607
127. Kunshenko BV, Il'nitzkii SO, Motnyak LA, Lyalin VV, Yagupolskii LM (1987) Zh. Org. Khim. 23:833
128. Wielgat J, Domagaza Z (1982) J. Fluor. Chem. 20:785
129. Lyalin VV, Grigorash RV, Alekseeva LA, Yagupolskii LM (1981) Zh. Org. Khim. 17:1774
130. Lyalin VV, Grigorash RV, Alekseeva LA, Yagupolskii LM (1984) Zh. Org. Khim. 20:846
131. Kunshenko BV, Ilnitskii SO, Motnyak LA, Lyalin VV, Burmakov AI, Yagupolskii LM: Bull. Soc. Chim. Fr. 1986:974
132. Muratov NN, Mokhamed NM, Kunshenko BV, Alekseeva LA, Yagupolskii LM (1986) Zh. Org. Khim. 22:964
133. Kunshenko BV, Muratov NN, Burmakov AI, Alekseeva LA, Yagupolskii LM (1983) Zh. Org. Khim. 19:1342

7 Fluorination of Organic Compounds with Fluorosulfuranes

Leonid Nikolaevich Markovskii and Valerii Efimovich Pashinnik

Institute of Organic Chemistry, Murmanskaya St, 5, 252094 Kiev, USSR

Contents

7.1 Introduction

The sulfur tetrafluoride derivatives, such as alkyl- and aryltrifluorosulfuranes, dialkylaminotrifluoro- and bis(dialkylamino)difluorosulfuranes, and polyfluoroalkoxytrifluorosulfuranes, are widely used in organic synthesis to substitute fluorine for oxygen in alcohols, aldehydes, ketones and acids, and for sulfur in thiocarbonyl compounds.

As opposed to sulfur tetrafluoride, most of the above fluorosulfuranes are liquids or crystalline substances, which allows one to conduct the reactions at atmospheric pressure and in glassware. Some derivatives of fluorosulfuranes, in particular, dialkylaminotrifluorosulfuranes, are more reactive than SF_4. All these factors provide high selectivity of fluorination, which is especially important in the case of polyfunctional compounds.

7.2 Fluorination with Alkyl- and Aryltrifluorosulfuranes

Alkyl- and aryltrifluorosulfuranes may be obtained by treatment of sulfur-containing compounds with fluorine, chlorine fluorides and iodine fluorides, and trifluoromethyl hypofluorite; by the electrochemical fluorination of sulfur-containing compounds; and by the addition of SF_4 to perfluoroolefins [1–11]. The most widely spread method for the preparation of these compounds is the general synthesis from dialkyl- and diaryl disulfides and silver difluoride suggested by Sheppard [12–14].

$$RSSR + 6AgF_2 \rightarrow 2RSF_3 + 6AgF$$

Thus treatment of diphenyl disulfide with AgF_2 gives phenyltrifluorosulfurane with a 60% yield, which is most frequently used for fluorination.

The reactions with aliphatic aldehydes and ketones are exothermal and are best controlled by using such solvents as dichloromethane or acetonitrile. The yields of the difluorides in these reactions do not exceed 50% [13].

$$AlkCHO + RSF_3 \rightarrow AlkCHF_2 + RSOF$$

$$AlkCOAlk' + RSF_3 \rightarrow AlkCF_2Alk' + RSOF$$

In mild conditions alkyl- and aryltrifluorosulfuranes also react with aromatic aldehydes. Ease of the reaction considerably depends on the electronic nature of the substituent in trifluorosulfurane. Thus the reaction of benzaldehyde with phenyltrifluorosulfurane at room temperature leads to benzal fluoride with a 80% yield [13]. A similar reaction with pentafluorophenyltrifluorosulfurane required heating to 100 °C, and the yield of benzal fluoride does not exceed 52% [14].

$$C_6H_5CHO + ArSF_3 \rightarrow C_6H_5CHF_2$$

$$Ar = C_6H_5 (80\%), C_6F_5 (52\%)$$

The aromatic ketones react with phenyltrifluorosulfurane in much more severe conditions, upon heating to 150 °C and in the presence of Lewis acids, for example, TiF_4 [13].

$$C_6H_5COC_6H_5 + PhSF_3 \xrightarrow{TiF_4} C_6H_5CF_2C_6H_5$$

The transformation of the carboxyl group to the trifluoromethyl one under the action of alkyl- and aryltrifluorosulfuranes occurs upon heating of the reagents to 120 to 150 °C in the presence of Lewis acids [13].

$$CH_3(CH_2)_5COOH + 2RSF_3 \rightarrow CH_3(CH_2)_5CF_3 + 2RSOF + HF$$

$$R = CF_3, C_6H_5$$

By contrast with similar reactions of SF_4 where such widespread Lewis acids as BF_3 and HF are used, these catalysts cannot be used for the reaction with phenyltrifluorosulfurane because of their high volatility, though good results have been obtained using a non-volatile Lewis acid—titanium tetrafluoride.

Dibasic acids such as acetylenedicarboxylic acid, also react with $PhSF_3$, with both carboxyl groups converted to the trifluoromethyl ones [15].

The relative reactivities of alkyl- and aryltrifluorosulfuranes and SF_4 in similar reactions have not been determined because of the different physical state of the reagents and the difficulty to compare the results. But the routes of substitution of oxygen by two fluorine atoms in the reactions of aldehydes and ketones with alkyl- and aryltrifluorosulfuranes are supposed to be the same as in the similar reactions of SF_4 [16].

7.3   Fluorination with Polyfluoroalkoxytrifluorosulfuranes

The subsitution of the hydroxy group in alcohols, and of the carbonyl group in aldehydes and ketones under the action of SF_4 is supposed to proceed via the stage of formation of alkoxytrifluorosulfuranes which rearrange with elimination of thionyl fluoride to form the respective fluoro-derivatives [16].

$$ROH \; + \; SF_4 \; \longrightarrow \; \left[\begin{array}{c} R\!-\!\!\!-\!O \\ \diagdown \; \mid \\ F\!-\!\!\!-SF_2 \end{array} \right] \; \longrightarrow \; RF \; + \; SOF_2$$

Recent synthesis of stable polyfluoroalkoxytrifluorosulfuranes and studies of their thermolysis allowed to establish some tendencies of their thermal transformations leading to fluorinated products. α,α,ω-Trihydropolyfluoroalkoxytrifluorosulfuranes were obtained by the reaction of alcohols of the general formula $H(CF_2)_nCH_2OH$ (n = 2, 4, 6) and SF_4 in the presence of potassium or sodium fluorides [17].

$$H(CF_2)_n CH_2 OH + SF_4 \xrightarrow[-70 \, to \, -50 \, °C]{KF \; or \; NaF} H(CF_2)_n CH_2 OSF_3$$

The thermal stability of these compounds considerably depends on the size of polyfluoroalkyl substituent, increasing from sulfurane with n = 2 to sulfurane with n = 4 and 6. The transformations of polyfluoroalkoxytrifluorosulfuranes have been found to depend on the thermolysis conditions. Prolonged keeping of alkoxytrifluorosulfurane with n = 6 at 5 °C leads to symmetrization products, and heating of it to 180 °C during 30 min affords fluoroalkane with a quantitative yield [17].

$$H(CF_2)_6 CH_2 OSF_3 \begin{array}{c} \xrightarrow{\;5\,°C\;} SF_4 + [H(CF_2)_6 CH_2 O]_4 S \\ \\ \xrightarrow[180\,°C]{} SOF_2 + H(CF_2)_6 CH_2 F \end{array}$$

Polyfluoroalkoxytrifluorosulfuranes may themselves behave as fluorinating agents. The reactions of 1,1,7-trihydrododecafluoroheptoxytrifluorosulfurane with carboxylic acids and their derivatives have been studied. The reaction route was shown to depend on the reaction temperature and the presence of Lewis acids in the reaction mixture. In the absence of Lewis acids, the reaction produces only acyl fluorides [18,19].

$$RCOOX + H(CF_2)_6 CH_2 OSF_3 \rightarrow RCOF + H(CF_2)_6 CH_2 OSOF + XF$$

$$R = CH_3, \; C_6H_5; \; X = H, \; Na, \; SiMe_3$$

In the presence of Lewis acids, the critical effect on the reaction route is produced by the temperature. At low temperatures, the reaction products are acyl fluorides, above 0 °C the reaction gives polyfluoroalkyl esters of these acids [19].

$$RCOOH + H(CF_2)_6CH_2OSF_3 \begin{cases} \xrightarrow{-40 \text{ to } -5\,°C} RCOF + H(CF_2)_6CH_2OSOF + HF \\ \xrightarrow{0\,°C} RCOOCH_2(CF_2)_6H + SOF_2 + HF \end{cases}$$

Polyfluoroalkoxytrifluorosulfuranes were shown to react with Lewis acids to form the polyfluoroalkoxydifluorosulfonium salt which is stable to $-5\,°C$, and above this temperature decomposes, liberating thionyl fluoride [19].

$$H(CF_2)_6CH_2OSF_3 + A \rightarrow [H(CF_2)_6CH_2OSF_2]^+AF^-$$

$$H(CF_2)_6CH_2O\overset{+}{S}F_2 \quad AF^- \begin{cases} \xrightarrow[-40 \text{ to } -5\,°C]{RCOOX} [RCOOSF_2OCH_2(CF_2)_6H] \\[4pt] \rightarrow RCOF + H(CF_2)_6CH_2OSOF \\[4pt] \xrightarrow[-SOF_2]{-5\,°C} [H(CF_2)_6\overset{+}{C}H_2 \quad AF^-] \end{cases}$$

$$\xrightarrow{RCOOX} RCOOCH_2(CF_2)_6H + XAF$$

$A = HF, BF_3, SbF_5$

7.4 Fluorination with Dialkylaminotrifluoro- and Bis(dialkylamino)difluorosulfuranes

Much more accessible fluorinating agents are dialkylaminotrifluorosulfuranes (DAST). Methods for their preparation are based on the reaction of SF_4 with secondary amines, dialkylaminotrimethylsilanes, tetraalkyldiamides of methylphosphonic acid, tetraalkyldiamides of sulfinic acid, dialkylamides of alkylsulfinic acid, etc. [20–30,32]. The most widespread method for the synthesis of dialkylaminotrifluorosulfuranes has become the reaction of SF_4 with dialkylaminotrimethylsilanes in a solvent at low temperatures [26–28, 32].

$$R_2NSiMe_3 + SF_4 \xrightarrow{-70 \text{ to } -50\,°C} R_2NSF_3 + Me_3SiF$$

$$R = Me, Et; \quad RR = -(CH_2)_4-, \ -(CH_2)_5-, \ -(CH_2)_2O(CH_2)_2-$$

DAST are mobile liquids distillable in vacuum, easily hydrolyzable and they decompose slowly upon storage. The most stable compound among the DAST studied is morpholinotrifluorosulfurane, it may be stored in the absence of air moisture for several years. Diethylaminotrifluorosulfurane, frequently used to fluorinate various types of compounds, is a commercially available product.

The high reactivity and selectivity of DAST allows one to use them to substitute fluorine for the carbonyl oxygen in aldehydes and ketones, sulfur in thiocarbonyl compounds, hydroxy group in alcohols, rather reactive chlorine in compounds of various types, and to transform the carboxyl group to the fluoroformyl one [26,27,31].

Aliphatic aldehydes and ketones react with DAST with evolution of heat. The yields of fluorination products exceed 55% [27,33–41].

$$AlkCHO + R_2NSF_3 \rightarrow AlkCHF_2 + R_2NSOF$$

$$Alk_2CO + R_2NSF_3 \rightarrow Alk_2CF_2 + R_2NSOF$$

If a molecule has several functional groups, the reaction with DAST results in their selective exchange for fluorine atoms. Thus in the fluorination of dispiro 3.1.3.1 decan-5,10-dione by an equimolar amount of DAST in CCl_4 at 25 °C for 30 h, only one keto group is substituted. Treatment of the resulting difluoro-derivative with the fluorinating agent in the same conditions results in the fluorination of another carbonyl group [37].

In the reaction of DAST with oxo carboxylic acids under mild conditions, the carbonyl oxygen is substituted by fluorines. In the case of ethyl 3-oxocarboxyla-tes, the fluorination proceeds at 20 °C during 50 h, the yield of the difluoride being 75% [39].

$$PhCH_2COCH_2COOEt \xrightarrow{\text{DAST}} PhCH_2CF_2CH_2COOEt$$

However in the case of the COOH group, in the reaction of 4-oxo carboxylic acids, there occurs fluorination and subsequent cyclization leading to fluorolactones with a high yield [40].

As a rule, the reactions of aldehydes and ketones with SF_4 are conducted in the presence of Lewis acids. Many aldehydes and ketones are unstable in acids, and DAST are therefore indispensable reagents for fluorination of such compounds.

It has been shown in the case of fluorination of pivalic aldehyde that the yield of the reaction products essentially depends on the solvent polarity. Thus in non-polar solvents such as $CFCl_3$ or pentane, the main product is gem-difluoride, the yields of the skeletal rearrangement products being no more than 13%. On passing to polar solvents, the yield of the difluoride decreases, and in diglyme it does not exceed 30% [27].

$$Me_3CCHO + Et_2NSF_3 \rightarrow Me_3CCHF_2 + CH_2{=}CMeCMeHF + FCMe_2CHFCH_3$$

Pentane	88%	2%	10%
Diglyme	30%	32%	38%

On the basis of these data, a scheme for fluorination of aldehydes with DAST has been suggested. The reaction is supposed to start with the electrophilic attack of the sulfurane sulfur atom at the carbonyl oxygen to form alkoxydialkylamino-sulfurane as an intermediate, whose dissociation to the ion pair and transformations of the latter, depending on conditions, lead to the end products [27].

$Me_3CCHO \quad + \quad Et_2NSF_3 \quad \longrightarrow \quad Me_3CCHF{-}OSF_2NEt_2$

$[Me_2\overset{+}{C}{-}CHFCH_3 \quad + \quad {}^-OSF_2NEt_2] \longleftarrow [Me_3C\overset{+}{C}HF \quad + \quad {}^-OSF_2NEt_2]$

$CH_2{=}CMeCHFCH_3 \quad Me_2CF{-}CHFMe \qquad Me_3CCHF_2$

a - non-polar solvent ; - polar non -basic solvent
c -polar basic solvent

Aromatic aldehydes react with DAST less vigorously; as a rule, heating of the reaction mixture to 50 to 60 °C is required. The yields of benzal fluorides are 65–75% [26,27,42,43].

$$ArCHO + Alk_2NSF_3 \rightarrow ArCHF_2 + Alk_2NSOF$$

The carbonyl group in fatty aromatic ketones is substituted by fluorine in much more severe conditions than in aromatic aldehydes. Thus the reaction of DAST with acetophenone proceeds with heating of a mixture of the reagents at 85 °C for 20 h [27].

$$C_6H_5COCH_3 + Et_2NSF_3 \rightarrow C_6H_5CF_2CH_3$$

A similar reaction of DAST with bromoacetophenone requires still more prolonged heating. It proceeds in benzene at 65 °C for 48 h [39].

$$C_6H_5COCH_2Br + Et_2NSF_3 \rightarrow C_6H_5CF_2CH_2Br$$

It was impossible to substitute the carbonyl group by fluorine in benzophenone. Thus mixing of morpholinotrifluorosulfurane with benzophenone does not result in a reaction even upon heating to 110 °C.

Acids of different types react with DAST, forming the respective acyl fluorides. The reaction is highly exothermal, the yields of acyl fluorides being 60–90% [26,44].

$$AcOH + Alk_2NSF_3 \rightarrow AcF + Alk_2NSOF + HF$$

$$Ac = AlkCO, ArCO, R_2PO, CF_3SO_2$$

The oxygen atom of the fluoroformyl group may not be substituted by fluorine even upon heating to 110 °C, and at higher temperatures DAST start to decompose.

However there is an example of substitution of the carbonyl group by fluorine by the reaction with DAST. Heating of an excess of DAST with benzoic acid at 80 °C for 20 h in diglyme gave benzotrifluoride with a 50% yield [42].

DAST may be used for the synthesis of acyl fluorides from the respective chlorides and to substitute fluorine for the reactive chlorine atoms in other compounds [31].

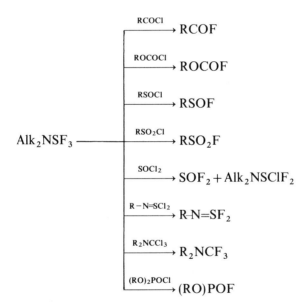

The yields of fluorination products in these reactions are 70–90%. Depending on the reactivity of the chloro-derivative and the method of isolation of the reaction products, the reactions are carried out by mixing the reagents in a solvent or without it.

The interaction of DAST with alcohols has been studied at length, and is a general method of substitution of the hydroxy group by fluorine. DAST in the fluorinations of alcohols has some advantages before other fluorinating agents used to substitute hydroxyl by fluorine, such as SF_4, the SeF_4–pyridine complex, α-fluoroalkylamines, HF, and HF-amine complexes [45–48].

This method is used to transform primary, secondary, and tertiary alcohols in mild conditions to the respective alkyl fluorides in high yields [27]. In the synthesis of low-boiling fluorides, a suitable solvent is diglyme, of high-boiling fluorides—pentane and dichloromethane.

The substitution of alcohol hydroxyl by fluorine under the action of various fluorinating agents is known to be accompanied by side reactions, such as rearrangements of various types and dehydration. Similar reactions are also characteristic for DAST. For instance, the reaction of i-BuOH with DAST leads to i-BuF and t-BuF in the ratio of 2:1, whereas its reaction with SeF_4-pyridine complex gives only the rearrangement product—$tert$-butyl fluoride [27].

$$(CH_3)_2CHCH_2OH + Et_2NSF_3 \rightarrow (CH_3)_2CHCH_2F + (CH_3)_3CF$$
$$49\% \qquad 21\%$$

The reaction of DAST with borneol and iso-borneol capable of forming stable carbocations, leads to the rearranged fluorides with 72 and 74% yields respectively [27].

The dehydration of the hydroxy-derivatives in the reactions with DAST occurs to a less extent than in fluorinations with other agents. Thus the reaction of DAST with cyclooctanol leads to cyclooctyl fluoride and cyclooctene in the ratio of 70:30. The reaction of cyclooctanol with the Yarovenko reagent (Et_2NCF_2CHClF) leads exclusively to the dehydration product—cyclooctene [27,47].

Isomeric 2-buten-1-ol and 3-buten-2-ol react with DAST to form a mixture of 3-fluoro-1-butene and 1-fluoro-2-butene. The formation of the same products from isomeric alcohols may be attributed to the ability of the latter to form the same carbocation, indicating the S_N1 mechanism of the reaction [27].

When one and the same molecule contains the hydroxyl and halogen, the latter remains intact in the reaction with DAST and does not produce any marked effect on the conditions of hydroxyl substitution [27,49–51].

$$Br(CH_2)_nCH_2OH + Et_2NSF_3 \rightarrow Br(CH_2)_nCH_2F$$

n = 1, 5, 7, 9

$$BrCH_2CH_2CHOHCH_2Br + Et_2NSF_3 \rightarrow BrCH_2CH_2CHFCH_2Br$$
$$76\%$$

When the starting molecule has both the alcoholic and phenolic hydroxyls, the latter remains intact in the reaction with DAST. Thus, *erythro*-3,4-bis(4-hydroxyphenyl)hexan-1-ol reacts with DAST at 20 °C forming *erythro*-1-fluoro-3,4-bis(4-hydroxyphenyl)hexane with a high yield [52].

The hydroxyl in fatty aromatic alcohols is substituted as easily as in ordinary aliphatic alcohols [27,53,54].

$$C_6H_5CH_2OH + Et_2NSF_3 \rightarrow C_6H_5CH_2F$$

$$C_6H_5CH_2CH_2OH + Et_2NSF_2 \rightarrow C_6H_5CH_2CH_2F$$

If a molecule of alcohol being fluorinated contains a sufficiently reactive carbonyl group, it also undergoes fluorination; these reactions are often accompanied by HF elimination and formation of unsaturated compounds [55,56]. For example, in the fluorination of 2-hydroxycyclohexanone, the main reaction product is difluorocyclohexene [55].

The fluorination of 7,12-dihydroxy-7(12)-methyl-7,12-dihydrobenz[a]an-thracenes is often accompanied by dehydrohalogenation specific to anthracene derivatives [57].

R^1 = Me R^2 = F; R^1 = F R^2 = Me

The ester and amide groups are stable against DAST. In the hydroxy-compounds containing these groups, only hydroxyl reacts. For example, in the reaction of ethyl lactate with DAST, fluorine is substituted for hydroxyl [27,58].

$$CH_3CHOHCOOEt + Et_2NSF_3 \rightarrow CH_3CHFCOOEt$$
$$78\%$$

In a similar way, with the phosphoryl group remaining intact, DAST react with α-hydroxyphosphonates. The reactions proceed in mild conditions, giving the respective α-fluoroalkyl phosphonates with good yields [59,60].

$$R'C_6H_4CR(OH)PO(OEt)_2 + Et_2NSF_3 \xrightarrow{CH_2Cl_2} R'C_6H_4CRFPO(OEt)_2$$

R=R'=H; R=H, R'=3-Cl, 4-Cl, 4-Me; R=Me, R'=H, 4-Cl

In a similar way DAST react with α-hydroxyphosphinates [60].

$$PhCH(OH)POPh_2 + Et_2NSF_3 \rightarrow PhCHFPOPh_2$$

However α-trimethylsilyloxyalkyl phosphonates do not react with DAST, and α-fluoroalkyl phosphonates may not be obtained in this way [60].

In contrast to this, trimethylsiloxyalkanes both of linear and iso-structure react in very mild conditions with DAST, forming fluoroalkanes in high yields. The reactions proceed with complete reversal of the configuration [61].

$$CH_3(CH_2)_5CH(OSiMe_3)CH_3 \xrightarrow{DAST} CH_3(CH_2)_5CHFCH_3$$
$$89\%$$

2-Octanol and 2-octyl tosylate are fluorinated in a similar way [61]. However, in the fluorination of bis(trimethylsiloxy)alkanes having Me_3SiO groups at the primary and tertiary carbon atoms, fluorine is substituted only for the siloxy group at the primary carbon atom.

$$\underset{\underset{OSiMe_3}{|}}{CH_3C(CH_3)(CH_2)_{11}OSiMe_3} \xrightarrow{DAST} \underset{\underset{OSiMe_3}{|}}{CH_3C(CH_3)(CH_2)_{11}F}$$
$$50\%$$

In some cases DAST shows a different reactivity towards the primary and secondary hydroxy groups. For example, the interaction of D-(*threo*)-1-(4-nitrophenyl)-2-phthalimido-1,3-propanediol with DAST at 20 °C leads to D-(*threo*)-1-(4-nitrophenyl)-2-phthalimido-3-fluoropropan-1-ol [62].

Mild and in many cases selective substitution of the hydroxy group by fluorine using DAST proved to be a convenient method for the synthesis of various physiologically active compounds [43,56,63–117]. DAST was used to synthesize a great number of 5α-androstanes. The fluorination proceeds selectively at room temperature, and only the hydroxy group is substituted by the fluorine atom, and the carbonyl group remains intact [65,66]. In the selective fluorination of the carbonyl group, the hydroxyl must be protected and then at approx. 80 °C the carbonyl group is smoothly transformed to the difluoromethylene one.

The route of the reaction of DAST with steroidal alcohols largely depends on the structure of the steroid [67]. If a molecule has the structure of 5-en-3-ol, the fluorination products are the respective 5-en-3-fluorosteroids. These reactions are supposed to form the intermediate carbocations stabilised by conjugation with the double bond. Substitution of the OH group by fluorine proceeds with preservation of configuration [67].

In the absence of the stabilising effect of double bond, the intermediate product is transformed by the S_N2 mechanism with reversal of configuration or via the carbocation, leading to rearrangements.

The reaction of DAST with cholestanol leads to 2-cholestene and 3α-fluorocholestene, and 3β-fluorocholestane is not formed [67].

If the reaction involves the *trans-diaxial* elimination, no fluorination products are formed. The main product in the reaction of 5α-pregnan-3α,17α-diol-20-one with DAST is the unsaturated compound formed as a result of HF elimination [67].

It is also impossible to carry out the fluorination of ergosterol and 7-dehydrositosterol. The reactions of these compounds with morpholinotrifluorosulfurane proceed with transformation of the position of double bonds and do not involve the substitution of the OH group in the 3-position by fluorine, giving the unsaturated compound [68–70].

The reaction is assumed to proceed via the intermediate formation of cation A, which is subsequently rearranged to the isomeric cation B stabilised by the proton elimination at C_{14} to form the end product.

If the 5,7-double bonds in ergosterol and 7-dehydrositosterol are protected, hydroxyl in these compounds may be easily substituted by fluorine.

The reactions of these compounds with morpholinotrifluorosulfurane give the difluoro-derivatives with approx. 50% yield, which after removal of protection from double bonds give 3β-fluoroergosterol and 3β-fluoro-7-dehydrositosterol. Thermal isomerisation of the latter gives 3β-fluorovitamins D_2 and D_5 [68,70].

DAST are used in the synthesis of fluorine-containing carbohydrates [79–96], e.g.:

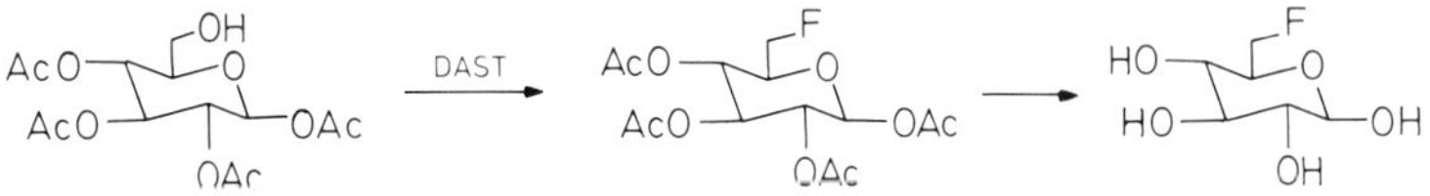

Recently, methods for the selective fluorination of various derivatives of α- and β-D-gluco- and mannopyranosides using DAST have been suggested. The fluorination of methyl-α-D-glucopyranoside with an excess of DAST without a solvent has been shown to proceed with substitution of two hydroxy groups to form methyl-4,6-didesoxy-4,6-difluoro-α-D-glucopyranoside. If the reaction is conducted in dichloromethane, selective monofluorination occurs to give methyl-6-fluoro-α-D-glucopyranoside with a good yield [87,88].

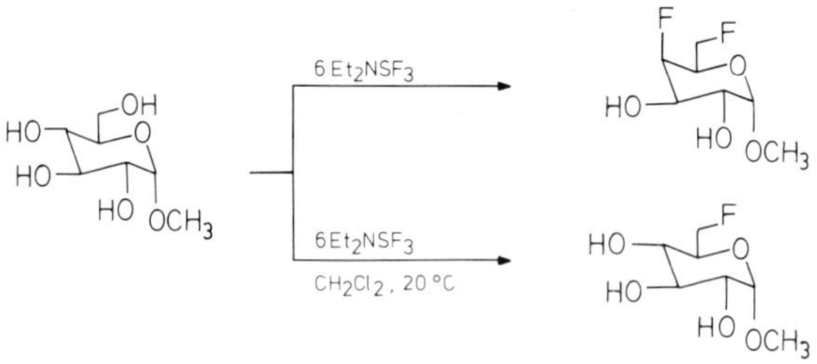

By contrast with α-glucosides, β-glucosides in dichloromethane and under the similar conditions as for α-glucosides, form the 3,6-didesoxy-3,6-difluoro-derivatives, but if the reaction time is reduced to 15 min, only the monofluoro-derivatives of β-glucosides may be obtained [88].

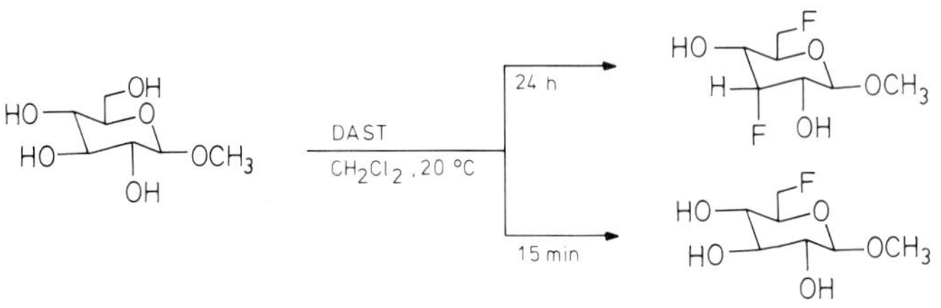

DAST find increasingly wide use in the synthesis of fluorine-containing prostaglandines and prostacyclines. Under the action of morpholinotrifluoro-sulfurane, the hydroxy group in prostaglandines A$_2$ and B$_2$ is substituted in mild conditions by fluorine. The ester and ketone groups remain intact [97–103].

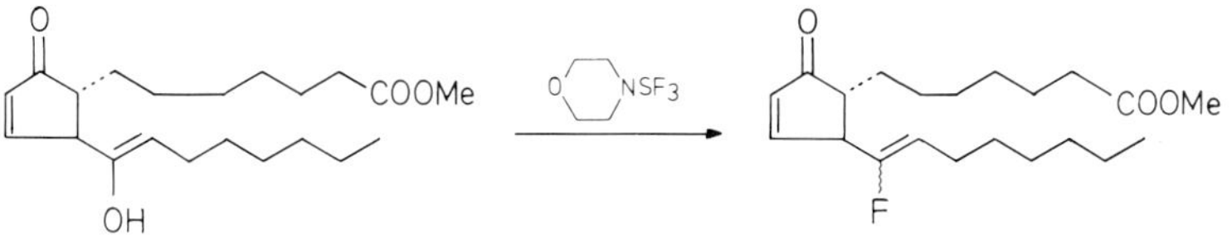

It is known that the steric position of fluorine often produces a critical effect on the type of biological activity. The attempted stereospecific substitution of the hydroxyl in prostaglandines by fluorine in the reaction of dimethylamino- and morpholinotrifluorosulfuranes resulted in the non-stereospecific formation of a mixture of α- and β-epimers (2:1) of C_{15}-fluorides. However treatment of methyl ester of prostaglandine A_2 with dimorpholinodifluorosulfurane leads to only one C_{15} epimer [101].

In the fluorination reactions, bis(dialkylamino)difluorosulfuranes are, as a rule, less active than DAST. For example, the reaction of 4-chlorobenzaldehyde with DAST starts at 60 °C and evolves much heat, whereas the similar reaction with dimorpholinodifluorosulfurane proceeds only under boiling in benzene for 1.5 h [29]. Bis(dialkylamino)difluorosulfuranes however may be effectively used to substitute the hydroxy group in highly reactive alcohols. The use of these reagents often allows to reduce side reactions—dehydration and isomerisation, as in the case of fluorination of 2-buten-1-ol and geraniol [27,104].

7.5 Preparations

1. A general procedure for the synthesis of DAST [26]

To a solution of SF_4 (0.12 mol) in 150 ml of anhydrous ether at -78 °C was added dropwise, with stirring, a solution of 0.1 mol of N,N-dialkyl-N-trimethyl-silylamine in 50 ml of ether. The reaction mixture was then gradually heated to 20 °C, an excess of SF_4 and ether was evaporated at reduced pressure, and the residue was fractionated. The yields of DAST were 60–70%. Diethylaminotrifluorosulfurane: b.p. 43 to 44 °C (12 mm). Morpholinotrifluorosulfurane: b.p. 41 to 42 °C (0.5 mm).

2. Fluorination with aryltrifluorosulfuranes [13]

Phenyltrifluorosulfurane (16.6 g, 0.1 mol) and benzaldehyde (10.6 g, 0.1 mol) were mixed in a 50 ml flask linked to a 45 mm column. Upon mixing of the reagents, an exothermal reaction starts. The reaction mixture was heated on an oil bath to 100 °C, then the pressure in the flask was reduced and benzal fluoride was distilled off (10.2 g, 80%) (b.p. 68 °C (80 mm)).

3. Fluorination of aldehydes and ketones [26]

DAST (0.1 mol) was mixed without solvent at 0 °C with 0.1 mol of carbonyl compound. The mixture was carefully heated to the start of the exothermal reaction, then heated at 60 °C for 15 min, whereupon it was dissolved in 20 ml of dichloromethane or CCl_4. The solution was poured into ice-water, stirred for 10 min, and the organic layer was separated. The mixture was dried and the solvent distilled off. The fluorination products were purified by distillation or recrystallisation. Yields are 55–75%.

4. Fluorination of alcohols [27]

To a solution of DAST in an inert solvent (CH_2Cl_2, pentane, diglyme, etc.) was added with cooling to -50 to $-78\,°C$ an equimolar amount of alcohol. The reaction mixture was then heated to room temperature. The exothermal reaction usually proceeds at a low temperature, but in some cases it occurs upon heating. The low-boiling fluorination products were distilled off at reduced pressure. The reaction mixture containing high-boiling fluorides was treated with water, the organic layer was separated and dried, and the solvent was distilled off. The fluorination products were purified by distillation, recrystallisation or column chromatography. Yields of the fluoro-derivatives in these reactions were 60–90%.

5. Fluorination of acids [26]

To a solution or suspension of 0.01 mol of acid in 30 ml of dry ether was added dropwise, with stirring and cooling with ice water, a solution of 0.01 mol of DAST in 10 ml of ether. The mixture was stirred at $20\,°C$ for 15 min. The acyl fluorides were isolated by filtration or fractional distillation. Yields were 60–90%.

6. Fluorination of acyl chlorides [31]

DAST was added dropwise, with stirring and cooling (in the case of exothermal reaction), to the equimolar amount of acyl chloride. The reaction mixture was stirred at $20\,°C$ for 15 min, and then at $60\,°C$ till the gaseous products ceased to evolve (about 30 min).

7.6   References

1. Tyczkowski EA, Bigelow LA (1953) J. Amer. Chem. Soc. 75:3523
2. Chambelain DL, Kharasch N (1955) J. Amer. Chem. Soc. 77:1041
3. Ratcliffe CT, Shreeve JM (1968) J. Amer. Chem. Soc. 90:5403
4. Sauer DT, Shreeve JM (1971) J. Fluor. Chem. 1:1
5. Abe T, Shreeve JM (1973/1974) J. Fluor. Chem. 3:17
6. Haran G, Sharp DWA (1973/1974) J. Fluor. Chem. 3:423
7. Sprenger GH, Cowley AH (1976) J. Fluor. Chem. 7:333
8. Middleton WJ, Howard EG, Sharkey WH (1965) J. Org. Chem. 30:1375
9. Denney DB, Denney DZ, Hsu YF (1973) J. Amer. Chem. Soc. 95:4064
10. US Pat 3456024 (1969); (1969) Chem. Abs. 71:70078
11. Rosenberg RM, Muetterties EL (1962) Inorg. Chem. 1:756
12. Sheppard WA (1960) J. Amer. Chem. Soc. 82.4751
13. Sheppard WA (1962) J. Amer. Chem. Soc. 84:3058
14. Sheppard WA, Foster SS (1973) J. Fluor. Chem. 2:53
15. Herkes FE, Simons HE (1975) J. Org. Chem. 40:420
16. Hasek WR, Smith WC, Engelhardt VA (1960) J. Amer. Chem. Soc. 82:543
17. Markovski LN, Bobkova LS, Pashinnik VE, Iksanova SV (1981) Zh. Org. Khim. 17:486
18. Markovski LN, Bobkova LS, Jaremenko VV, Pashinnik VE (1983) Zh. Org. Khim. 19:1632

19. Pashinnik VE, Tovstenko VI, Bobkova LS, Markovski LN (1985) Zh. Org. Khim. 21:2072
20. Demitras GG, Kent RA, MacDiarmid AG (1964) Chem. Ind. 38:1712
21. Demitras GG, MacDiarmid AG (1967) Inorg. Chem. 6:1903
22. Halasz SP, Glemser O (1970) Chem. Ber. 103:594
23. Halasz SP, Glemser O (1971) Chem. Ber. 104:1247
24. Gibson JA, Ibbott DG, Janzen AF (1973) Can. J. Chem. 51:3203
25. Brown DH, Crosbie KD, Darragh JI, Ross DS, Sharp DWA: J. Chem. Soc. (A) 1970:914
26. Markovski LN, Pashinnik VE, Kirsanov AV: Synthesis 1973:787
27. Middleton WJ (1975) J. Org. Chem. 40:574
28. Middleton WJ, Bingham EM (1977) Org. Synth. 57:50
29. Markovski LN, Pashinnik VE, Kirsanova NA (1975) Zh. Org. Khim. 11:74
30. Markovski LN, Pashinnik VE, Kirsanova NA (1976) Zh. Org. Khim. 12:965
31. Markovski LN, Pashinnik VE: Synthesis 1975:801
32. Braun C, Dell W, Sasse HE, Ziegler ML (1979) Z. anorg. allg. Chem. 450:139
33. Adcock W, Gupta BD, Khor-Thong-Chak (1976) Aust. J. Chem. 29:2571
34. US Pat 4416822 (1983)
35. Swiss Pat 616433 (1980)
36. Siegemund G (1979) Lieb. Ann. Chem. 9:1280
37. Sharts CM, McKee ME, Steed RF, Shellhamer DF, Greeley AC, Green RC, Sprague LG (1979) J. Fluor. Chem. 14:351
38. Mursakulov IG, Samoshin VV, Binnatov RV, Pashinnik VE, Povolotzkii NI, Zefirov NS (1983) Zh. Org. Khim. 19:1336
39. Buss CW, Coe PL, Tatlow JC (1986) J. Fluor. Chem. 34:83
40. Patrick TB, Poon YE (1984) Tetrahedron Lett. 25:1019
41. May JA, Sartorelli AC (1979) J. Med. Chem. 22:971
42. US Pat 3914265 (1975); (1976) Chem. Abs. 84:42635
43. US Pat 3950329 (1976); (1976) Chem. Abs. 85:78142
44. Radchenko OA, Il'chenko AJ, Yagupolskii LM (1980) Zh. Org. Khim. 16:863
45. US Pat 2980740 (1961); (1961) Chem. Abs. 55:23342
46. Olah GA, Nojima M, Kerekes J (1974) J. Amer. Chem. Soc. 96:925
47. Jarovenko NN, Raksha MA (1959) Zh. Obshch. Khim. 29:2159
48. Olah GA, Nojima M, Kerekes J: Synthesis 1973:786
49. Cavalho JF, Prestwich GD (1984) J. Org. Chem. 49:1251
50. Olah GA, Singh BP, Liang G (1984) J. Org. Chem. 49:2922
51. Saito K, Digenis GA, Hawi AA, Chaney J (1987) J. Fluor. Chem. 35:663
52. Goswami R, Harsy SG, Heiman DF, Katzenellenbogen JA (1980) J. Med. Chem. 23:1002
53. Middleton WJ, Ringham EM (1977) Org. Synth. 57:72
54. Ishikawa N, Kitazume T, Takaoka A (1979) J. Synth. Org. Chem. Jap. 37:606
55. US Pat 4112815 (1980); (1980) Chem. Abs. 93:239789
56. Cross BE, Simpsom JC: J. Chem. Res. 1980:118
57. Newman MS, Khanna JM (1979) J. Org. Chem. 44:866
58. Esfahani M, Cavanaugh SR, Preffer P (1981) Biochem. Biophys. Res. Commun. 101:306
59. Blackburn GM, Parratt MJ: J. Chem. Soc. Chem. Commun. 1983:886
60. Blackburn GM, Kent DE: J. Chem. Soc. Perkin Trans. I 1986:913

61. Asai T, Yasuda A, Matsumura Y, Kato M, Uchida K (1986) Repts. Res. Lab. Asachi Glass Co. 36:49
62. US Pat 4235892 (1980); (1981) Chem. Abs. 94:139433
63. Cross BE, Erasmuson A, Filipone P: J. Chem. Soc. Perkin Trans. I 1981:1293
64. Cross BE, Erasmuson A: J. Chem. Soc. Chem. Commun. 1987:1013
65. Bird TGC, Felsky G, Fredericks PM, Jones ERH, Meakin GD: J. Chem. Res. 1979:388
66. Bird TGC, Fredericks PM: J. Chem. Soc. Chem. Commun. 1979:65
67. Rosen S, Faust J, Ben-Yakov H (1979) Tetrahedron Lett. 20:1823
68. Jakhimovich RI, Fursaeva NF, Pashinnik VE: Khim. prirod. soed. 1980:580
69. Jakhimovich RI, Fursaeva NF, Baum VK, Valinietze MU, Apyhovskaya LI, Nekrasova NB, Kovalev VE (1981) Khim. Farm. Zh. 15:75
70. Jakhimovich RI, Fursaeva NF, Pashinnik VE: Khim. prirod. soed. 1985:102
71. Robays MV, Busson R, Vanderhaeghe H: J. Chem. Soc. Perkin Trans. I 1986:251
72. Mann J, Pietrzak B: J. Chem. Soc. Perkin Trans. I 1987:385
73. Biollaz M, Kalvoda J (1977) Helv. Chim. Acta 60:2703
74. Prestwich GD, Shieh HM, Gayen AK (1983) Steroids 41:79
75. Ger. Offen. 2632550 (1977); (1977) Chem. Abs. 87:23616
76. Yang SS, Dorn CP, Jones H (1977) Tetrahedron Lett. 27:2315
77. Napoli JI, Fivizzani MA, Schnoes HK, Deluca HF (1979) Biochem. 18:1614
78. Kobayashi Y, Taguchi T (1985) J. Synth. Org. Chem. Jap. 43:1073
79. Card PJ (1985) J. Carbohyd. Chem. 4:451
80. Sharma M, Korytnyk W: Tetrahedron Lett. 1977:573
81. Sharma M, Korytnyk W (1980) Carbohyd. Res. 83:163
82. Sufrin JR, Bernacki RS, Morin MS, Korytnyk W (1980) J. Med. Chem. 23:143
83. US Pat 4284764 (1981); (1982) Chem. Abs. 96:7032
84. Somawardhana CW (1981) Carbohyd. Res. 94:14
85. Tewson TJ, Welch MJ (1978) J. Org. Chem. 43:1090
86. Klemm GH, Kaufman RJ, Sighu RS (1982) Tetrahedron Lett. 23:2927
87. Card PJ (1983) J. Org. Chem. 48:393
88. Card PJ, Reddy GS (1983) J. Org. Chem. 48:4734
89. Albert R, Dax K, Katzenbeisser U, Sterk H, Stutz AE (1985) J. Carbohyd. Chem. 4:521
90. Druekhammer DG, Wong Chi Huey (1985) J. Org. Chem. 50:5912
91. Kovac P, Yah HJC (1986) J. Carbohyd. Chem. 5:497
92. Kovac P (1986) Carbohyd. Res. 153:168
93. Faghih R, Escribano FC, Castillon S, Garcia J, Lukacs G, Olesker A, Trang Ton That (1986) J. Org. Chem. 51:4558
94. Kovac P, Yeh HJC, Jung GL, Glademans CPJ (1986) J. Carbohyd. Chem. 5:497
95. Street IP, Withers SG (1986) Can. J. Chem. 64:1400
96. Binder TP, Robyt JE (1986) Carbohyd. Res. 147:149
97. Bezuglov VV, Bergel'son LD (1980) Dokl. Akad. Nauk SSSR 250:468
98. Shevchenko VP, Myasoedov NF, Bezuglov VV, Bergel'son LD (1981) Bioorg. Khim. 7:448
99. Bezuglov VV, Bergel'son LD (1979) Bioorg. Khim. 5:1531
100. Bezuglov VV, Serkov IV, Gafurov RG, Llerena EM, Pashinnik VE, Markovski LN, Bergel'son LD (1984) Dokl. Akad. Nauk SSSR 279:378
101. Pashinnik VE, Markovski LN, Bezuglov VV, Serkov IV, Gafurov RG, Bergel'son LD (1984) in: Tezisy 2nd Vses. Soveshch., Ufa, USSR

102. Bezuglov VV (1986) in: Tezisy Vses. Symp. Tallin, USSR
103. Serkov IV, Gafurov RG, Pashinnik VE, Tovstenko VI, Markovskii LN, Bezuglov VV, Bergel'son LD (1986) in: Tezisy Vses. Symp. Tallin, USSR
104. Poulter CD, Wittins PL, Plummer TL (1981) J. Org. Chem. 46:1532
105. Asato AE, Liu RSH (1986) Tetrahedron Lett. 27:3337
106. US Pat 4188345 (1980)
107. US Pat 4226787 (1980)
108. US Pat 4229358 (1980)
109. US Pat 4263214 (1981)
110. US Pat 4229357 (1980)
111. US Pat 4230627 (1980)
112. ·US Pat 4224230 (1980)
113. Gani D, Hitchcock PB, Young DW: J. Chem. Soc. Chem. Commun. 1983:898
114. Tsushima T, Sato T, Tsuji T (1980) Tetrahedron Lett. 21:3591
115. An Seung-Ho, Bobek M (1986) Tetrahedron Lett. 27:3219
116. Biggadike K, Borthwick AD, Evans D, Exall AM, Kirk BE, Roberts SM, Stephenson L, Young P, Slawin AM, Williams DJ: J. Chem. Soc. Chem. Commun. 1987: 251
117. US Pat 4397783 (1983); (1983) Chem. Abs. 99:175473

Subject Index